Symbolic Computation
Applications to Scientific Computing

Frontiers in Applied Mathematics

Managing Editors
for Practical Computing Series

W.M. Coughran, Jr.
AT&T Bell Laboratories
Murray Hill, New Jersey

Eric Grosse
AT&T Bell Laboratories
Murray Hill, New Jersey

Frontiers in Applied Mathematics is a series that presents new mathematical or computational approaches to significant scientific problems. Beginning with Volume 4, the series reflects a change in both philosophy and format. Each volume focuses on a broad application of general interest to applied mathematicians as well as engineers and other scientists.

This unique series will advance the development of applied mathematics through the rapid publication of short, inexpensive books that lie on the cutting edge of research.

Frontiers in Applied Mathematics

Symbolic Computation
Applications to Scientific Computing

Edited by **Robert Grossman**
University of Illinois at Chicago

Society for Industrial and Applied Mathematics

siam   Philadelphia 1989

Library of Congress Cataloging-in-Publication Data

Symbolic computation : applications to scientific computing / [edited
    by] Robert Grossman.
            p.   cm. -- (Frontiers in applied mathematics ; 5)
    The outgrowth of talks at the NASA-Ames Workshop on the Use of
Symbolic Methods to Solve Algebraic and Geometric Problems Arising
in Engineering, which took place on January 15 and 16, 1987, at the
NASA-Ames Research Center in Moffett Field, California"--Pref.
    Includes bibliographical references.
    ISBN 0-89871-239-4
    1. Engineering mathematics--Data processing--Congresses.
2. Object-oriented programming--Congresses. I. Grossman, Robert,
1957-   .   II. Society for Industrial and Applied Mathematics.
III. Ames Research Center. IV. Series.
TA331.S95   1989
620'.001'51--dc20   89-19701
                                                               CIP

Contents

Contributors

Harold Abelson, Artificial Intelligence Laboratory, Massachusetts Institute of Technology, Cambridge, Massachusetts 02139

O. Akhrif, Electrical Engineering Department and Systems Research Center, University of Maryland, College Park, Maryland 20742

G. L. Blankenship, Electrical Engineering Department and Systems Research Center, University of Maryland, College Park, Maryland 20742

Philip Colella, Department of Mechanical Engineering, University of California at Berkeley, Berkeley, California 94720

Richard Fateman, Department of Computer Science, University of California at Berkeley, Berkeley, California 94720

Matthew Grayson, Department of Mathematics, University of California at San Diego, La Jolla, California 92093

Robert Grossman, Department of Mathematics, Statistics, and Computer Science, University of Illinois at Chicago, Chicago, Illinois 60680

Paul N. Hilfinger, Department of Electrical Engineering and Computer Science, University of California at Berkeley, Berkeley, California 94720

P.S. Krishnaprasad, Electrical Engineering Department and Systems Research Center, University of Maryland, College Park, Maryland 20742

Richard H. Rand, Department of Theoretical and Applied Mechanics and Center for Applied Mathematics, Cornell University, Ithaca, New York 14853

N. Sreenath, Systems Engineering Department, Case Western Reserve University, Cleveland, Ohio 44121

Gerald Jay Sussman, Department of Electrical Engineering and Computer Science, Massachusetts Institute of Technology, Cambridge, Massachusetts 02139

Preface

The papers in this volume are the outgrowth of talks at the NASA-Ames Workshop on the Use of Symbolic Methods to Solve Algebraic and Geometric Problems Arising in Engineering, which took place on January 15 and 16, 1987 at the NASA-Ames Research Center in Moffett Field, California.

Although symbolic computation has been an important tool for researchers for some time now, a typical reaction after working with a computer algebra package is often, "this is very useful, but if only the software could directly support the structures and objects that I use in my own work." The speakers at the workshop described research which illustrated how objects such as domains, maps, vector fields, Lie algebras, difference operators, perturbation series, dynamical systems, Poisson manifolds, and flows could be manipulated symbolically by designing the proper software. The resulting software was then used in a variety of areas, including: chaotic systems, fluid dynamics, nonlinear control systems, and robotics.

Most of the papers had to deal, in one form or another, with the question of how a computer system can effectively compute not only functions and expressions, but also complicated objects such as operators. The first chapter in the volume is a brief survey of some of the issues that arise in a computer algebra system that must support operators. It concludes with a brief introduction to the other chapters.

Research in the boundary between control theory and computer science is in a time of rapid change. A recent white paper on control theory concludes that there is a need for research in the interface between symbolic and numeric computation and control theory[1]:

---

[1]W.H. Fleming, ed., *Report of the Panel on Future Directions in Control Theory: A Mathematical Perspective*, SIAM, Philadelphia, p. 49, 1988.

In recent years considerable interest has been generated in the area of symbolic computation for control. Although there is some research on combined symbolic and numeric computation, this area of research has a substantial potential for both academic research and engineering development. The interface of symbolic and numerical computing with computer graphics is a research area that could have direct impact on mathematical analysis, the development of CAD design tools, and even hardware capabilities.

The workshop not only bore witness to the truth of this statement, but also indicated that the impact of control theory on computer science promises to be just as exciting.

Beginning in 1986, the NASA-Ames Research Center under the direction of Dr. George Meyer has sponsored a series of workshops entitled "Artificial Intelligence and Control Theory." The format for each workshop has been similar: over a two-day period approximately one hundred engineers, computer scientists, and mathematicians gather to hear talks, which last an hour and half, and are followed by a half hour of questions and discussion. Here are the titles and dates of the first three workshops.

Research Issues in Robotics, January, 1986

The Use of Symbolic Methods to Solve Algebraic and Geometric Problems Arising in Engineering, January, 1987

Workshop on AI and Discrete Event Control Systems, July, 1988

These workshops have had an important impact on research. The support and direction of Dr. Meyer and NASA-Ames are greatly appreciated.

Robert Grossman
Chicago

Chapter 1

Computer Algebra and Operators

Richard Fateman and Robert Grossman

1 The Symbolic Computation of Operators

After defining the two expansions

$$\exp(A) = \sum_{i=0}^{N} A^i/i!$$

$$\log(1+A) = \sum_{i=1}^{N} (-1)^{i+1} A^i/i,$$

a computer algebra system such as Macsyma or Maple will quickly compute

$$\log(\exp(A)\exp(B)) = A + B + \mathrm{O}(N+1).$$

Here $\mathrm{O}(N+1)$ denotes terms containing a product of $N+1$ or more A's and/or B's. This computation depends crucially upon the fact that $AB = BA$; for objects for which this not true, certain correction terms enter. For example, if A and B are matrices, then in general $AB \neq BA$ and

$$\log(\exp(A)\exp(B)) = A + B - \frac{1}{2}AB + \frac{1}{2}BA + \ldots.$$

The symbolic computation of operator expansions such as these differs in a number of ways from the symbolic computation of expressions in commuting variables. The papers in this volume consider various aspects of such computations. In this introductory chapter, we first discuss some of the capabilities that prove useful when performing computer algebra computations involving operators. These capabilities may be broadly divided into three areas: the algebraic manipulation of expressions from the algebra generated by operators; the algebraic manipulation of the actions of the operators upon other mathematical objects; and the development of appropriate normal forms and simplification algorithms for operators and their actions. We then conclude the introduction by giving a little background and a brief description of the problems considered by the authors in this collection.

1.1 Algebra of Operators

Let $E_1, E_2, \ldots, E_M$ denote operators and α a number. Then $E_1 + E_2$, $E_2 E_1$, and αE_1 are all operators of the same type. That is, a finite set of operators naturally generate an algebra of operators under addition and composition. Let $\mathbf{R}\{E_1, \ldots, E_M\}$ denote this algebra. This is just the free associative algebra over $\mathbf{R}$ generated by the symbols $E_1, \ldots, E_M$. The first requirement of a computer algebra system for operators, then, is that it support the algebraic operations of addition and composition of operators.

The first issue raised is how to represent operators and operations on them in a context which has already usurped most of the notation for an algebra of expressions. Is it possible to use some multiplication operator (typically "*") for operators, or should one use another notation? Maple [2] uses "@" and Macsyma [4] uses "." for composition, but juxtaposition for application. Mathematica [6] takes no special notice of this, but allows the use of juxtaposition for multiplication. (It thereby maps f(a) into the same expression as a*f; if you intend to convey the notion $f(a)$, you type f[a].) In fact, the notation and semantics of operators have at best been a patch added on to conventional general-purpose algebra systems; by contrast, the most effective computer implementation of operator algebras has been quite application specific, as, for example, in the case of manipulation of perturbation series.

Yet the need for combining operator notation with the generality of widely available computer algebra systems dictates that we seriously address the representation and manipulation of operators. A good test for any system is to take some simple and familiar concepts and see if they can be naturally expressed and manipulated. For example, an operator that expresses "differentiation with respect to x" should be natural. Extensions to represent the 2nd or the nth derivative should be natural. Mixed derivatives (with respect to x and y) should be possible: a natural operation might be the interchange of orders of differentiation, and the combination of common variable "powers." Evaluation of the derivative at a point, say the origin (the un-obvious $f'(0)$), is a challenge [2]. Because algebra systems are generally built to be extensible by procedure definition, data-type extension, transformation-rule (or pattern-match) extension, and aggregations of data, it is natural for one to hope that new operators can be defined, and their significant properties encoded, in a natural form.

There is a tension between expressiveness and precision. In different contexts, we have different expectations. Should the algebra system be expected to treat identical notations in different ways? Consider the notation $(L + a)(y)$. This might mean (say if L is a derivative operator) $dy/dx + ay$. Yet in other circumstances we might prefer that a constant a be an operator equivalent to a function which returns a, regardless of its argument. In that case, $(L + a)(y) = Ly + a$. If an algebra system requires unambiguous notation, it may be unnatural in nearly all mathematical contexts. For the two variations above, Maple would use `<L@x+a*x|x>@y` and `<L@x+a|x>@y`, respectively while Macsyma would require a declaration `declare(a,opscalar)` then `(L+a)y` for the first and would use `(L+constop(a))y` for the latter. A proposal and implementation for Macsyma by T. H. Einwohner [3] would use the notation `L+a*I |` `y` to mean $L(y) + a * y$. (Unfortunately, the use of "|" is at odds with the Maple convention.) Einwohner's design is somewhat more regular in reflecting the duality between expressions involving operators (which can be quite arbitrary), and the results of applying the operator expression to arguments. The use of parameters is an important component of the design. For example, `f(a)|y` is an alternative notation for `f(y,a)`. Finally, he does not further overload the use of "." as noncommutative multiplication by using it for operator composition.

Another issue is how to best structure the data and the manipulative algorithms for handling expressions from free associative algebras. These issues have been examined since the earliest days of computer algebra systems, and are perhaps the aspects of the computer algebra of operators which can be attacked entirely by conventional means; this usually constitutes a mapping into algebraic tree representations, where most algorithms can be treated as tree-traversals (as described in any text on data structures). On the other hand, truly efficient manipulation may dictate more compact representations, including so-called string-polynomial manipulations, matrix encodings, or other schemes that have been devised for hand manipulation.

1.2   Actions of Operators

The usefulness of operators to scientists and engineers derives not from their properties as abstract algebraic objects but rather from their interesting actions on other mathematical objects. For example, matrices act as linear transformations on vector spaces, vector fields act as derivations on spaces of functions, the physicist's raising and lowering operators act on the space of polynomials, the algebra generated by shift operators acts on the space of sequences, and the algebra of maps acts on the space of domains.

This brings us to the second major requirement on a computer algebra system for operators. They must support the action of operators on objects from the natural domains of definition of the operators (and presumably the natural domains of the algebra system). For example, having formed the matrix expression $E = A + A^2/2! + \ldots + A^4/4!$, it is useful to be able to apply E to vectors. The merging of the two notations leads to a complexity and variety of actions that is probably the single most important barrier to satisfactory implementations of operators in computer algebra systems. The operator E above certainly looks like a polynomial in A; for some purposes it can best be treated as a polynomial; in other contexts as in section 1, it certainly is not equivalent.

There is a more subtle, yet related issue. There is no natural mathematical notation to specify the various interlocking operations possible on the mixed domain of operators and operands. For example, consider an operator F, and its inverse which we will conventionally write as F^{-1} (even though it may have rather little to do with $1/F$) and an expression $F\ F^{-1}\ x$. It is natural for us to apply simplification and "cancel" the operators, yielding x. It is not so obvious to us or to the computer, though, whether in more complex situations to apply or simplify. Is FFx computed faster if F is "squared" first? Or is $F(Fx)$, where the parentheses suggest some ordering, preferable?

Sometimes partial simplification is useful. Consider a definition of a new operator $F := < \sum_{i=1}^{n} x \mid x,\ n >$ where we use Maple's notation to indicate that x and n are the formal names used for two actual "parameters" to the operator. If G is another operator (presumably operating on positive integers) then $F(G, 3)$ is $G(1) + G(2) + G(3)$. Consider $< F(I, k) \mid k >$, where I is an identity operator. At what time (if ever) would a computer algebra system realize that this is equivalent to $< k(k+1)/2 \mid k >$?

What kind of syntax do we supply for the user to define such a simplification? Furthermore, how do we deal with an action of a particular operator on a domain which is not supported? For example, should the syntax describing the action of shift operators acting on sequences be the same as the syntax describing the action of vector fields as derivations on the space of functions?

How can the implementation of operators as algebraic objects be best merged with the implementation of operators as objects acting on other objects in the computer algebra system?

It appears that linguistically, two approaches are possible, and these are not addressed in Maple or Macsyma. One is to require unambiguous specification of operations such as operator simplification (so they would occur on command), a distinct notation for application of operators, and an explicit translation (back and forth) from operator to functional notation.

Another approach is to use the domains of objects (themselves possibly other operators) to disambiguate the meanings of operators, at least whenever possible. This may require fairly sophisticated pattern-matching support which checks the types of arguments. An example used by Gonnet [2] illustrates this. Consider the expression $a \times D \times D \times y$. If we assume D is an operator, each of the three "multiplications" may be a different concept. The first is symbolic multiplication by a constant. The second is composition, and the

third is application. Yet we were able to disambiguate this by look-
ing at the types of the "arguments" to ×. A constant on the left,
a, means traditional multiplication; a non-operator on the right, y,
means that the operator to the left is being applied to it. A multiplication between two operators means composition. Note that it
would be an error to simply work from the right to the left, applying as we go, although for this expression it might work. Consider
a non-integrable form z, and the integration operator D^{-1}. Then
$D\ D^{-1}\ z$ could not be simplified, because the application of the integration operator would not "work" (unless D were specially coded
to "unwrap" an integral).

There are many open questions, and it appears to us that the best
process for making headway in the introduction of operator calculi
in algebraic manipulation is to explore serious applications, ones
which are challenges for humans and computers. Without further
experience such as we see in this volume of papers, it is too easy to
make false starts and empty promises. We do not know, for example,
if it make sense for the language describing the action of matrices
on vectors to be the same as the language describing the action of
differential operators on the space of functions.

While Maple proposes a methodology primarily based on an extension of procedures, with some original notation for operators,
Macsyma uses (for the most part) already existing notation for noncommutative multiplication; we expect that users of the Mathematica system will primarily use pattern matching and notation which
looks fundamentally like functional application. Each approach has
its advantages and advocates.

1.3   Normal Forms and Simplification

Data structures and algorithms to manipulate operators depend crucially on the action of the operators upon objects from the natural
domains of definition of the operators. Normal forms for expressions built from matrices are probably not the best normal forms for
expressions built from maps. Questions about normal forms, simplification, and efficient evaluation of operator expressions are by and
large open.

It appears that another significant area of application of a computer algebra system is the manipulation of operator expressions to

produce simplified or if possible normal forms for the various types and combinations of supported operators by the computer algebra system. This may involve transformation of operator expressions to a well-supported domain (typically polynomials), or collections of rules which drive expressions algorithmically or heuristically, toward equivalent but simpler forms.

Given the state of the art, it seems inevitable that a successful general-purpose system will have to provide some facility for users to implement their own normal forms and simplification algorithms for more specialized types of operators and actions.

2 Examples of Operators and their Domains

In this section we give brief descriptions of the computer algebra computations that arise when working with various operators and their actions. The papers in this collection give careful and complete descriptions of how the authors dealt with some of the issues mentioned above.

2.1 Perturbation Theory

A classical problem in perturbation theory is to compute approximate solutions to differential equations containing small parameters. Consider van der Pol's equation

$$\frac{d^2 x}{dt^2} + x - \epsilon(1 - x^2)\frac{dx}{dt} = 0,$$

where $\epsilon \neq 0$ is a small parameter. The starting point is to expand a putative solution $t \to x(t)$ in a power series in ϵ

$$x(t) = x_0(t) + \epsilon x_1(t) + \epsilon^2 x_2(t) + \ldots,$$

and then substitute this series into the original differential equation to obtain a sequence of differential equations (one for each power of ϵ) for the unknown coefficient functions $x_i(t)$. These auxiliary differential equations have the property that the ith equation involves only the coefficient functions $x_0(t)$, $\ldots$, $x_i(t)$, so that the sequence of differential equations can be solved recursively. In the example above, the equations are

$$\frac{d^2}{dt^2} x_0(t) + x_0(t) = 0$$

$$\frac{d^2}{dt^2} x_1(t) + x_1(t) = \left(\frac{d}{dt} x_0(t)\right)(1 - x_0{}^2(t))$$

$$\vdots$$

There are several ways of approaching these types of computations. Letting $\dot{x}(t) = y(t)$, van der Pol's equation can be written as the first order system

$$\dot{x}(t) = y(t)$$

$$\dot{y}(t) = -x(t) + \epsilon(1 - x^2(t))y(t).$$

Let E_1 denote the vector field $y\partial/\partial x - x\partial/\partial y$ and let E_2 denote the vector field $(1 - x^2)y\partial/\partial y$. Then van der Pol's equation becomes

$$\dot{z}(t) = (E_1 + \epsilon E_2)(z(t)),$$

where $z(t) = (x(t), y(t))$. Notice that $E_2 E_1 \neq E_1 E_2$.

Let T_n denote the operator which acts upon functions $t \to z(t)$ and returns the first nth terms in a power series expansion in ϵ of the function. Then the ith auxiliary equation is equivalent to the system of equations which is the coefficient of ϵ^i in the expansion of

$$(d/dt - (E_1 + \epsilon E_2))T_n(z(t)) = 0,$$

for n sufficiently large. From this point of view, perturbation theory is concerned with the symbolic algebra of noncommutative operators such as E_1 and E_2 acting on the domain of power series expansions of the form

$$f(x, y) = f_0(x, y) + \epsilon f_1(x, y) + \epsilon^2 f_2(x, y) + \cdots.$$

A different but related point of view is used by R. Rand in his contribution "Perturbation Methods and Computer Algebra." Rand describes a computer algebra system built using Macsyma, which in a systematic fashion can perform serious perturbation computations, including the computation of normal forms and center manifold reductions.

2.2 Finite Difference Operators and Domains

Consider the heat equation in a bounded region Ω of the Euclidean plane

$$\frac{\partial u}{\partial t} = \frac{\partial^2 u}{\partial x^2} + \frac{\partial^2 u}{\partial y^2}, \quad \text{for } x \in \Omega \text{ and } t \geq 0.$$

To compute the temperature $u(x,t)$ numerically using finite differences, we need to discretize the domain Ω, the function $u(x,t)$, and the differential operator

$$\frac{\partial}{\partial t} - \frac{\partial^2}{\partial x^2} - \frac{\partial^2}{\partial y^2}.$$

This can be done in many ways. Let D denote a finite mesh of points

$$\begin{aligned} x_i &= x_0 + i\triangle x \quad \text{for } i = 0, \ldots, I \\ y_j &= y_0 + j\triangle y \quad \text{for } j = 0, \ldots, J \end{aligned}$$

covering the region Ω, and let $u_{i,j}^n$ denote the temperature at time $n\triangle t$ at the mesh point $(x_0 + i\triangle x, y_0 + j\triangle y)$.

With the shift operators

$$\begin{aligned} E_x(k)u_{i,j}^n &= u_{i-k,j}^n \\ E_y(k)u_{i,j}^n &= u_{i,j-k}^n, \end{aligned}$$

we can define the difference operators

$$\begin{aligned} \delta_x^2 u_{i,j}^n &= \left(E_x(-1) - 2E_x(0) + E_x(1)\right) u_{i,j}^n \\ &= u_{i+1,j}^n - 2u_{i,j}^n + u_{i-1,j}^n \\ \delta_y^2 u_{i,j}^n &= \left(E_y(-1) - 2E_y(0) + E_y(1)\right) u_{i,j}^n \\ &= u_{i,j+1}^n - 2u_{i,j}^n + u_{i,j-1}^n, \end{aligned}$$

and compute the temperature $u_{i,j}^{n+1}$ given the temperature $u_{i,j}^n$ implicitly using the scheme

$$u_{i,j}^{n+1} = u_{i,j}^n + \frac{1}{2}\left(\frac{\triangle t}{\triangle x^2}\right)\left(\delta_x^2 u_{i,j}^{n+1} + \delta_x^2 u_{i,j}^n + \delta_y^2 u_{i,j}^{n+1} + \delta_y^2 u_{i,j}^n\right).$$

Notice that the basic objects are: domains, such as D; functions defined on domains, such as $u_{i,j}^n$; and operators defined on the functions, such as $E_x(k)$ and δ_x^2. Also, notice that the latter two objects can both be thought of as maps: a function on a domain can be thought of as a map from the domain to the range of the function; an operator on functions can be thought of as a map from the space of functions on domains to itself. In a later chapter in this volume, "FIDIL: A Language for Scientific Programming" by P. Hilfinger and P. Colella, the language FIDIL (FInite DIfference Language) is introduced. This language makes domains and maps basic data types and provides for the efficient computation of objects built from them, making the translation of standard numerical algorithms into programming statments very simple.

Related work is contained in [1].

2.3   Automatic Derivation of Dynamical Equations

The time evolution of a mechanical system consisting of a system of rigid bodies joined together with massless ball and socket joints can be quite complicated. In fact, even to write down the correct equations of motion can be difficult. It would be very useful if a program could automatically derive the equations of motion of this type of mechanical system. Similarly, an electrical circuit consisting of resistors, capacitors, and voltage sources can exhibit interesting dynamical behavior. It is an interesting problem to write a program whose input consists of a description of a mechanical or electrial system and whose output consists of the differential equations governing the time evolution of the state variables of the system.

A description of the system would include the following:

System parameters. System parameters must be defined and specified. For example, the moment of inertia of a body is defined to be $\int_{body} Q \cdot Q \, dm(Q)$, where $m(Q)$ is the mass distribution of the body. To describe the system, the mass distributions and topology of the connections of the rigid bodies must be given, and the moments of inertia must be computed.

Dynamical variables. The dynamical variables must be defined. For example, the rotation matrix $B(t)$ of a rigid body, which specifies how the body is turning in space, must be defined. Using this, the angular velocity $\Omega(t)$ of the body can be defined via $\Omega(t) = \dot{B}(t)B^{-1}(t)$. As a another example, Newton's Law $F(t) = dp(t)/dt$ defines the force $F(t)$ acting on a body in terms of the momenta $p(t)$ of the body. Both these example are typical, in the sense that dynamical variables are often defined by differentiating other dynamical variables. Notice that this gives rise to differential identities satisfied by the dynamical variables.

Algebraic relations. The dynamical variables not only satisfy differential relations, but typically algebraic relations determined by the geometry of the particular system. For example, if $r_1(t)$ and $r_2(t)$ denote the positions of the center of mass of bodies 1 and 2, respectively, $r(t)$ denotes the position of the center of mass of the ensemble of the two bodies, $d_1(t)$ and $d_2(t)$ their initial displacements, and $B_1(t)$ and $B_2(t)$ the rotation matrices describing the rotation of the bodies, then

$$r(t) = B_1(t)d_1(t) + B_2(t)d_2(t).$$

State variables. Because of the algebraic and differential relations satisfied by the dynamical variables, it is possible to select some of the dynamical variables, called state variables, and to construct all of the other dynamical variables from these. For example, for two bodies, only the position and velocity of the two bodies and their center of mass need be specified. Alternately, the position and momenta of the two bodies, and of their center of mass, can be specified.

Evolution of state variables. Finally, differential equations giving the time evolution of the state variables need to be given, together with the corresponding initial conditions. For example, the initial positions and velocities of each of the bodies and of each of the joints must be specified in order for the dynamical equations giving the time evolution of the state variables to be integrated.

The contribution "Multibody Simulation in an Object Oriented Programming Environment" by N. Sreenath and P.S. Krishnaprasad undertakes the automatic derivation of the equations of motion for systems of coupled rigid bodies in the plane, while the contribution "The Dynamicist's Workbench: I, Automatic Preparation of Numerical Experiments" by H. Abelson and G. J. Sussman treats electrical networks consisting of resistors, capacitors, and voltage sources. Both papers start with a description of the system, and by using a variety of symbolic and symbolic/numeric techniques eventually compute numerical simulations of the differential equations which the programs automatically compute.

The papers treat the problem of finding state variables differently. The paper by Abelson and Sussman finds state variables using a symbolic Newton iteration to eliminate relations among dynamical variables. The paper by Sreenath and Krishnaprasad uses symmetries of the mechanical system and a technique from geometric mechanics called reduction to derive equations of motion on a smaller dimensional phase space consisting of state variables.

Studies such as these make heavy use of symbolic computation of operators. For example, the dynamical equations of a system of coupled rigid bodies can be written as $\dot{F}(t) = \{F, H\}$, where H is the Hamiltonian of the system, and both F and H evolve on the appropriate phase space P. Here $\{\cdot, \cdot\}$ is a noncommutative operator defined on

$$\mathbf{C}^{\infty}(P) \times \mathbf{C}^{\infty}(P) \to \mathbf{C}^{\infty}(P).$$

2.4   Lie Brackets and Vector Field Systems

Recall that a smooth vector field $F(x)$ defined in a neighborhood of the origin in $\mathbf{R}^N$ is of the form

$$F(x) = \sum_{i=1}^{N} a_i(x) \frac{\partial}{\partial x_i},$$

where $a_1(x), \ldots, a_N(x)$ are smooth functions defined in a neighborhood of the origin. The Lie bracket of two vector fields $F(x)$ and $G(x)$ is defined to be

$$[F, G] = F(x)G(x) - G(x)F(x).$$

It is a fundamental property of Lie brackets that although they appear to be second-order differential operators they are in fact first-order differential operators because of the cancellation of the second-order partial derivatives. In other words, the Lie bracket of two vector fields is a vector field. It therefore makes sense to form iterated Lie brackets such as $[[F, G], G]$.

Observe that the Lie bracket of two vector fields is a measure of how far from commuting the two vector fields are. For example, the Lie brackets of the coodinate vector fields $F_i = \partial / \partial x_i$ are all zero, indicating that these vector fields all commute with each other. In other words, $F_i F_j = F_j F_i$. On the other hand, the Lie bracket of the vector fields $F_1 = x_2 \partial/\partial x_1$ and $F_2 = \partial/\partial x_2$ is equal to

$$[F_2, F_1] = \partial/\partial x_1.$$

Let $F(x)$ and $G_1(x), \ldots, G_m(x)$ denote smooth vector fields defined in a neighborhood of the origin in $\mathbf{R}^N$. It turns out that the local behavior of the nonlinear control system

$$\dot{x}(t) = F(x(t)) + \sum_{i=1}^{m} u_i(t) G_i(x(t)), \qquad x(0) = x_0 \in \mathbf{R}^N,$$

is determined by the algebraic properties of iterated Lie brackets of the F and G's. This is analogous to the fact that the local behavior of a smooth function is determined by its Taylor coefficients. Because of this fact, questions about the dynamical behavior of control trajectories can be reduced to symbolic questions about the algebraic properties of the noncommutative operators $F, G_1, \ldots, G_m$ acting on the domain $\mathbf{C}^\infty(\mathbf{R}^N)$ of smooth functions on $\mathbf{R}^N$.

The contribution "Symbolic Computations in Differential Geometry Applied to Nonlinear Control Systems" by O. Akhrif and G. L. Blankenship describes a software package written in the computer algebra system Macsyma, which through the symbolic computation of the proper Lie bracket expressions, can analyze system theoretic properties of a nonlinear control system, such as feedback equivalence, left-invertibility, or the design of control laws.

The contribution "Vector Fields and Nilpotent Lie Algebras" by M. Grayson and R. Grossman considers conditions on the Lie brackets which guarantee that the trajectories of a vector field system can be explicitly integrated in closed form. In general this cannot be expected and the game is to find as large a class of such systems as possible. The paper gives some simple examples of such systems which were computed used the computer algebra system Maple.

Acknowledgments

Richard Fateman's research was supported in part by NSF grant number CCR-8812843; DARPA order #4871, contract N00039-84-C-0089, the IBM Corporation; the State of California MICRO program. Robert Grossman was supported in part by grant NAG2-513 from the NASA-Ames Research Center.

References

[1] Balaban, David, "The Automatic Programming Project," Lawrence Livermore National Laboratory Technical Report, 1987.

[2] Gonnet, Gaston H., "An implementation of operators for symbolic algebra systems," in *Proc. ACM SYMSAC '86 Conf.*, Waterloo, Ontario, Canada, pp. 239–243, 1986.

[3] Einwohner, Theodore H. "A Vaxima Implementation of Operator Calculus," Lawrence Livermore National Laboratory Computer Science Report TS88-60, 1988.

[4] Golden, Jeffery P.,"An operator algebra for Macsyma," *Proc. ACM SYMSAC '86 Conf.*, Waterloo, Ontario, Canada, pp. 244–246, 1986.

[5] Watt, Stephen M. and Dora, Jean Della, "Algebra snapshot: linear ordinary differential operators," *Scratchpad II Newsletter*, Vol. 1, pp. 14–18, 1986.

[6] Wolfram, Stephen, *Mathematica: A System for Doing Mathematics by Computer*, Addison-Wesley, Reading, Mass., 1988.

Chapter 2

The Dynamicist's Workbench: I
Automatic Preparation of Numerical Experiments

Harold Abelson and Gerald Jay Sussman

Before computers, exploring the behavior of a dynamical system was tedious, often requiring great skill to make the computations tractable. In 1801 it took a Gauss to compute the orbits of Ceres and Pallas; with a computer, anyone can do it. But experimental dynamics is still tedious: An investigator repeatedly selects interesting values for parameters and initial conditions, sets up numerical computations, decides when each run is completed, and classifies the results, decomposing the system's parameter space into regions of qualitatively different behavior. Even with powerful numerical computers, experimental dynamics typically requires significant human effort to set up simulations and relies upon human judgment to choose parameter values that are "interesting,"to determine when each simulation run is "complete," and to decide when two behaviors are "qualitatively different." Much of this work, however, can be automated.

Our goal is to produce a *dynamicist's workbench*—a computer system that could, in principle, write many of the published papers that describe the behaviors of particular dynamical systems. Such a system must be able to set up and evolve numerical simulations. It must exploit algebraic and geometric constraints to determine when a simulation will produce no new interesting behavior, to classify trajectories and to recognize the bifurcations of critical sets. In this paper we describe a portion of our system that deals with the setting up and execution of numerical simulations.[1] This part of the workbench includes a spectrum of computational tools—numerical methods, symbolic algebra, and semantic constraints (such as dimensions). These tools are designed so that combined methods, tailored to particular problems, can be constructed on the fly. One can use symbolic algebra to automatically generate numerical procedures, one can use domain-specific constraints to guide algebraic derivations and to avoid complexity, and one can use numerical methods to identify and verify qualitative properties of systems.

We illustrate these ideas in the context of a few dynamical systems initially formulated as electrical networks. Section 1 presents a language for describing electrical networks. From these descriptions, the workbench generates the algebraic constraints and other information needed to support analysis. Section 2 illustrates how the workbench evolves the state of a dynamical system by automatically compiling a procedure to compute the system derivative, and combining this with an appropriate numerical integrator composed from primitive integrators and one of a number of strategies for adaptive step-size control. These system-derivative procedures generated by the workbench may incorporate iteration schemes when the state-variable derivatives cannot be expressed in closed form. Section 3 describes the automatic compilation of procedures that compute the frequency response of linear systems. This requires substantial symbolic manipulation, which is made tractable by using semantic markers to guide the algebra. In section 4 we demonstrate how one can explore the complex dynamics of the driven van der Pol oscillator. Here the workbench automatically compiles numerical procedures for finding periodic orbits and for tracking them as the system parameters vary.

1   Algebraic Environments

Information about a dynamical system to be analyzed using the workbench is organized in an *algebraic environment*, which is a structure that maintains algebraic constraints among variables. A variable may be annotated with semantic markers describing its role in

---

[1] Ken Yip's Ph.D. thesis, in progress, shows how a program can classify trajectories and bifurcations and use the results to choose promising values for parameters and initial conditions.

a dynamical system. For example, some variables may be parameters of the system, while others may be state variables. One variable may name a quantity with dimensions of length while another may have dimensions of capacitance. The workbench performs symbolic algebra on expressions in these variables, using the semantic markers to guide the algebraic manipulations. Algebraic environments can be constructed from explicitly specified constraints and annotations. Alternatively, one can employ a special-purpose language tailored for constructing algebraic environments from descriptions of dynamical systems such as electrical circuits or signal-flow graphs.

The examples in this paper are drawn from electrical circuit theory. They are initially formulated in terms of a network-description language. The network language contains a few predefined parts corresponding to the simplest electrical elements: *resistor*, *capacitor*, *inductor*, *voltage-source*, and *current-source*. There are also two primitives in terms of which all these elements can be defined: the *branch* and the *constraint*. A branch defines an arbitrary two-terminal device with its associated quantities, and a constraint establishes a relationship among quantities. Using branches and constraints, one can also define elements such as op-amps and nonlinear resistors.

In the network language, compound networks are constructed by connecting together (primitive or non-primitive) parts. Any compound network, once defined, can be used as a part in constructing a still more complex network. The expression `define-network` is used to name newly-constructed networks. Here is a definition of the Twin-T circuit shown in Figure 1 consisting of three resistors, three capacitors, and a voltage source:

```
(define-network twin-t () (n1 n2 n3 n4)
  (parts (s voltage-source (n+ n3) (n- gnd))
         (r1 resistor (n+ n3) (n- n2))
         (r2 resistor (n+ n2) (n- n4))
         (r3 resistor (n+ n1) (n- gnd))
         (c1 capacitor (n+ n3) (n- n1))
         (c2 capacitor (n+ n1) (n- n4))
         (c3 capacitor (n+ n2) (n- gnd))))
```

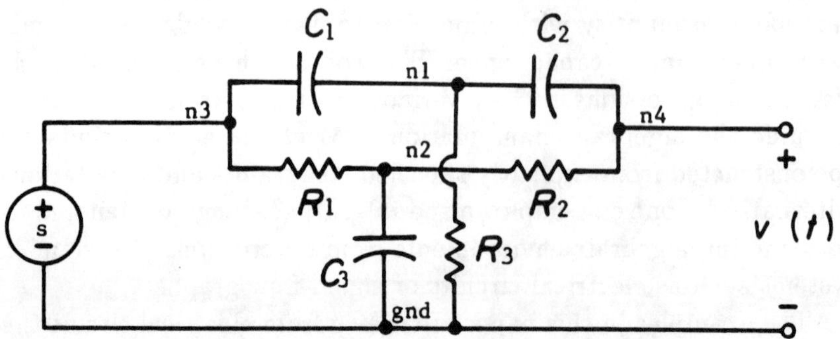

Figure 1. The Twin-T network is a third-order linear system that is often used as a notch filter in audio applications.

The general form of a network definition is

```
(define-network type arguments nodes
  (parts part1
         part2
         ...
         partn))
```

Type is the name for the new type of network being defined. **Arguments** and **nodes** are each a list of information that may be specified when a network of this type is used (see below). Each of the **part** entries consists of a name for the part, a specification of the type of the part, and specifications for the arguments and nodes of the part.

The nodes in a network definition describe the topology of how the parts are connected. This is accomplished by showing correspondence between the names of the nodes of the parts and the names of the nodes of the network being defined. In the case of the **twin-t** definition, there are four nodes: n1, n2, n3, and n4. The definition of a capacitor as a primitive element includes two nodes, n+ and n−. The **twin-t** subpart specification

```
(c3 capacitor (n+ n2) (n- gnd))
```

stipulates that c3 is a capacitor whose **n+** node is identified with the circuit node n2 and whose n− node is identified with the node

gnd, which is an additional external node defined for all circuits. Observe that the names of nodes defined for a part are local to the part. Thus, the capacitors and resistors each refer to their own two nodes as **n+** and **n-**, and in each case these are identified with different nodes of the Twin-T network. In a similar way, if we were to define a network that used a **twin-t** as a part, the appropriate part specification would indicate how the nodes **n1–n4** are to be identified with nodes defined in the larger network. In general, each pair of names in a part specification matches a name defined locally for the part with an object from the context in which the part is embedded.

Declarations and Constraints

A network definition can be interpreted to generate algebraic constraints and semantic marker declarations for construction of an instance of the type described by the definition. We can construct an algebraic environment containing an instance of the prototype Twin-T network as follows:

```
(define twin-t-inst (make-algebraic-instance twin-t))
```

Some algebraic variables represent device parameters (such as the resistance of a resistor). Others represent circuit variables (such as node potentials, branch voltages, and terminal currents). Some circuit variables (such as the voltage across a capacitor) may be state variables of the system. Among the declarations created by executing the **twin-t** description are

```
(r.r2 parameter resistance)
(c.c1 parameter capacitance)
((v.r3 t) circuit-variable voltage)
((v.c2 t) state-variable voltage)
```

Thus, **r.r2** is a parameter with the dimensions of a resistance, and **c.c1** is a parameter with the dimensions of a capacitance. **V.r3** is a function of time, is a circuit-variable, and has the dimensions of a voltage. **V.c2** is a function of time, is a state-variable of the capacitor[2] and has the dimensions of a voltage.

Here are some of the constraints that are developed by interpreting the **twin-t** definition:

---

[2] **V.c2** is a state-variable of the capacitor considered in isolation, but it is not necessarily a state-variable of the network in which the capacitor is embedded.

```
(fact140 (= (+ (- 0 (i.r1 t)) (i.r2 t) (i.c3 t)) 0))

(fact155 (= (v.c3 t) (- (e.n2 t) (e.gnd t))))

(fact156 (= (i.c3 t) (* c.c3 (rate (v.c3 t) ($ t)))))
```

Fact140 is Kirchhoff's Current Law at node n2. Fact155 stipulates that the voltage across capacitor c3 is the difference of the potentials at the nodes n2 and gnd (Kirchhoff's Voltage Law). Fact156 is the constituent relation for the capacitor c3, asserting that the current is the product of the capacitance and the time-derivative of the voltage.[3]

Declarations and constraints are derived ultimately from descriptions of primitive network elements, such as the following description of a capacitor:

```
(define-network capacitor
  ((c parameter capacitance)
   (v state-variable voltage)
   (i circuit-variable current))
  (n+ n-)
  (primitive-element
   (constraints '(= ,v (- ,(potential n+) ,(potential n-)))
                '(= ,i (* ,c ,(rate v))))
   (current-from-node '(,n+ ,i) '(,n- (- 0 ,i)))))
```

The prototype capacitor has a parameter C with the dimensions of a capacitance, a state variable v with the dimensions of a voltage, and a circuit-variable i with the dimensions of a current. There are two constraints—Kirchhoff's Voltage Law and the constituent relation $i = C\,dv/dt$. The capacitor also answers the question "What is the current entering you from node m?" When parts are connected to form networks, Kirchhoff's Current Law is enforced by asserting, at each node, the constraint that the sum of the currents entering each part from that node is zero.

The circuit language is only one example of a language for translating domain-specific descriptions into algebraic constraints. Another example (also provided in the workbench although not discussed here) is a language for describing the block-diagram systems used in signal processing. One could also imagine a similar language

---

[3]The dollar-sign syntax in fact156 indicates that the derivative (rate) is to be taken with respect to t.

for describing mechanisms constructed from rods, cams, and gears. In each case, in moving from the problem domain to the algebra, one needs to capture not only the algebraic equations but the semantic markers as well. We shall see below how this semantic information is used.

2   Evolving Time-Domain Behavior

Given a dynamical system specified in terms of an algebraic environment, the workbench automatically generates procedures that support the simulation of the system. Some of these procedures are numerical. Others are higher-order "generators" that will be specialized when the simulation is run. These procedures are automatically combined with input and graphical output routines to generate simulation programs. If we instruct the workbench to prepare a time-domain simulation, we will be asked to specify an initial state and values for the parameters. The workbench will use these values to evolve the corresponding time behavior, and can report the values of any variable contained in the algebraic environment, or of any algebraic expression in these variables. Figure 2 shows three graphs produced by the workbench for the `twin-t` network described above.

2.1   Generating a System Derivative

Given an algebraic environment containing constraints and semantic markers, the workbench's first step in producing a time-domain simulation is to express the derivatives of the state variables in terms of the parameters and the state variables, performing algebraic manipulation to eliminate intermediate variables as necessary.[4] In any particular circuit, the state variables of the individual parts, such as the voltages of capacitors, may be dependent. The workbench recognizes such dependencies and automatically chooses an independent set of state variables.

If the system's state vector is $\mathbf{y}$ then the result of this manipu-

---

[4]In nonlinear systems, solving explicitly for the derivatives may be beyond the capabilities of the algebraic manipulator. Section 2.4 describes how the workbench constructs iterative methods for dealing with such situations.

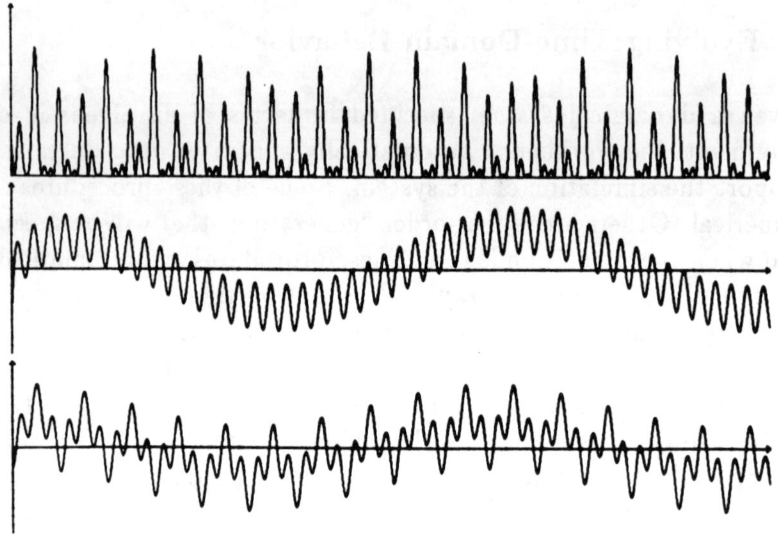

Figure 2. This is the time-domain behavior of the Twin-T network as evolved
by the workbench. The parameters are $C_1 = C_2 = 0.1$, $C_3 = 0.2$, $R_1 = R_2 = 1$,
$R_2 = 0.5$, and the source voltage is $\cos t - \cos 10t - \cos 30t$. The initial state
(at $t = 0$) is **v.c1** = **v.c2** = **v.c3** = 0. The horizontal scale is $[0, 10]$ seconds.
The bottom trace shows the source voltage—a superposition of three sinusoids.
The vertical scale is $[-4, 4]$ volts. The middle graph show the potential at node
n4—with this choice of parameters, the Twin-T network is behaving as a notch
filter, suppressing the middle-frequency component of the input. The vertical
scale here is $[-2, 2]$ volts. The top graph shows the power dissipated in resistor
r1. The vertical scale here is $[0, 5]$ watts.

lation is a first-order system of the form:

$$\frac{d\mathbf{y}}{dt} = \mathbf{f}(t, \mathbf{y})$$

The resulting algebraic expressions for the components of the system derivative, $\mathbf{f}$, are compiled to form a *system-derivative generator* procedure. The generator takes as arguments numerical values for the system parameters and produces a *system-derivative* procedure, which takes a system state vector as argument and produces a differential state (a vector that when multiplied by an increment of time is an increment of state). The system-derivative procedure will be passed to an *integration driver* that returns a procedure, which given an initial state, evolves the system numerically. Other expressions are also compiled into procedures that compute values from a system state vector.

To eliminate variables, the solver repeatedly chooses an unused equation, picks a variable from the class of variables to be eliminated, and solves the equation for that variable. The resulting value is bound to the variable in the algebraic environment, so that subsequent evaluations of expressions with respect to that algebraic environment will return values that have no instances of the eliminated variables. Backsubstitutions are thus automatically subsumed by the evaluation process. Although for particular classes of equations, such as linear systems, there are vastly more efficient algorithms for producing the system derivative, we have chosen this elimination strategy because it can easily tolerate a few nonlinear relations, and it degrades gracefully when it cannot make progress.

This elimination strategy is simple in outline, but in practice it is important to carefully order the sequence of eliminations to minimize the size of intermediate expressions. Treating operator and function applications as variables also requires care. Consider the two equations $dx/dt = dy/dt$ and $y = x + z$ where we regard dx/dt and dy/dt as variables. If we first eliminate dx/dt (reducing it to dy/dt), then eliminate y (reducing it to $x+z$), and then attempt to evaluate dx/dt, we will not have performed a complete elimination, because dx/dt will reduce to $dx/dt + dz/dt$.

Example: The System Derivative for the Twin-T Network

For the Twin-T network the workbench produces the following expressions for the derivatives of the three state variables:

```
(rate (v.c3 t) ($ t)) = (/ (+ (* r.r1 (strength.s t))
                               (* -1 r.r1 (v.c1 t))
                               (* -1 r.r1 (v.c2 t))
                               (* -1 r.r1 (v.c3 t))
                               (* r.r2 (strength.s t))
                               (* -1 r.r2 (v.c3 t)))
                            (* c.c3 r.r1 r.r2))

(rate (v.c2 t) ($ t)) = (/ (+ (strength.s t)
                               (* -1 (v.c1 t))
                               (* -1 (v.c2 t))
                               (* -1 (v.c3 t)))
                            (* c.c2 r.r2))

(rate (v.c1 t) ($ t)) = (/ (+ (* r.r2 (strength.s t))
                               (* -1 r.r2 (v.c1 t))
                               (* r.r3 (strength.s t))
                               (* -1 r.r3 (v.c1 t))
                               (* -1 r.r3 (v.c2 t))
                               (* -1 r.r3 (v.c3 t)))
                            (* c.c1 r.r2 r.r3))
```

These expressions are compiled to form the system-derivative
generator procedure shown in Figure 3. The generator takes as arguments the **twin-t** network's seven declared parameters—three capacitances, three resistances, and the strength of the source. (The
strength of the source is a procedure that computes a function of
time.) The result returned by the generator is a procedure whose
single argument is a state vector with four components—time, and
the voltages across the three capacitors. Notice that the workbench's
expression-compiler has performed a bit of common-subexpression
removal to make derivative computation more efficient and to pre-
vent multiple evaluation of the **strength.s** procedure.

2.2 Compiling Auxiliary Expressions

In addition to generating algebraic expressions for the system derivatives, the workbench includes a general-purpose *algebraic evaluator*
that can be called to evaluate algebraic expressions relative to a
given algebraic environment. For instance, we can ask for the power
(product of current and voltage) in the Twin-T network's resistor
r1:

```
==> (algebra-value '(* (v.r1 t) (i.r1 t)) (twin-t-inst 'time-domain))

(/ (+ (* (strength.s t) (strength.s t))
      (* -2 (strength.s t) v.c3)
      (* v.c3 v.c3))
   r.r1)
```

We can verify (as Tellegen's theorem implies) that the sum of the
powers into all the elements in the network is zero:

```
==> (algebra-value '(+ (* (v.r1 t) (i.r1 t))
                       (* (v.r2 t) (i.r2 t))
                       (* (v.r3 t) (i.r3 t))
                       (* (v.c1 t) (i.c1 t))
                       (* (v.c2 t) (i.c2 t))
                       (* (v.c3 t) (i.c3 t))
                       (* (v.s t) (i.s t)))
                 (twin-t-inst 'time-domain))
0
```

We can also instruct the workbench to compile numerical procedures to compute various expressions, such as the powers into each
of the resistors, as functions of the state and the parameters:

```
==> (compile-time-expressions '((* (v.r1 t) (i.r1 t))
                                (* (v.r2 t) (i.r2 t))
                                (* (v.r3 t) (i.r3 t)))
                              (twin-t-inst 'time-domain))
```

The procedure compiled for these expressions is shown in Figure 4.
The top graph in Figure 2 was produced by plotting the values of
each first component in the sequence of triples generated using this
procedure.

2.3   Generating Methods of Integration

To evolve a dynamical system, a system derivative is combined with
an *integration driver* to produce a procedure which, when called
with an initial state, evolves the state numerically. Here is a typical
integration driver:

```
(define (system-integrator system-derivative max-h method)
  (let ((integrator (method system-derivative)))
    ;; integrator : state-and-step ---> state-and-step
```

```
(lambda (c.c3 c.c2 c.c1 r.r3 r.r2 r.r1 strength.s)
  (lambda (*state*)
    (let ((t (vector-ref *state* 0))
          (v.c3 (vector-ref *state* 1))
          (v.c2 (vector-ref *state* 2))
          (v.c1 (vector-ref *state* 3)))
      (let ((g2 (strength.s t)))
        (let ((g6 (* g2 r.r2)) (g4 (* -1 v.c1))
              (g3 (* -1 v.c3)) (g5 (* -1 v.c2))
              (g1 (* -1 r.r1)))
          (vector 1
                  (/ (+ g6
                        (* g1 v.c1)
                        (* g1 v.c2)
                        (* g1 v.c3)
                        (* g2 r.r1)
                        (* g3 r.r2))
                     (* c.c3 r.r1 r.r2))
                  (/ (+ g2 g3 g4 g5)
                     (* c.c2 r.r2))
                  (/ (+ g6
                        (* g2 r.r3)
                        (* g3 r.r3)
                        (* g4 r.r2)
                        (* g4 r.r3)
                        (* g5 r.r3))
                     (* c.c1 r.r2 r.r3)))))))))
```

Figure 3. The system-derivative generator procedure compiled for the Twin-T
network takes as arguments a set of parameters and returns as its value a procedure that takes a state vector and returns a differential state. Each of the
four components of the returned differential state is the time derivative of the
corresponding component of the state vector.

```
(lambda (c.c3 c.c2 c.c1 r.r3 r.r2 r.r1 strength.s)
  (lambda (*state*)
    (let ((t (vector-ref *state* 0))
          (v.c3 (vector-ref *state* 1))
          (v.c2 (vector-ref *state* 2))
          (v.c1 (vector-ref *state* 3)))
      (let ((g7 (strength.s t)))
        (let ((g9 (* -2 g7))
              (g12 (* g7 g7)))
          (let ((g11 (+ (* g9 v.c1) (* v.c1 v.c1)))
                (g8 (+ g12 (* g9 v.c3) (* v.c3 v.c3)))
                (g10 (* 2 v.c1)))
            (list (/ g8 r.r1)
                  (/ (+ g11
                        g8
                        (* 2 v.c2 v.c3)
                        (* g10 v.c2)
                        (* g10 v.c3)
                        (* g9 v.c2)
                        (* v.c2 v.c2))
                     r.r2)
                  (/ (+ g11 g12)
                     r.r3)))))))))
```

Figure 4: This is the procedure generator compiled to compute the powers dissipated by each of the three resistors in the Twin-T network. Note the extensive common-subexpression removal performed by the workbench here.

```
(define (next state-and-step)
  (output (state-part state-and-step))
  (let ((new-state-and-step (integrator state-and-step)))
    (next (make-state-and-step
            (state-part new-state-and-step)
            (min (step-part new-state-and-step) max-h)))))
  next))
```

System-integrator takes as arguments a system derivative, a maximum step-size max-h, and a general method of integration. The method is applied to the system derivative, producing an integrator which, given a data structure that contains a state and a step-size, integrates for one step. In order to admit integrators with adaptive step-size control, integrator is structured to return not only the next state, but also a predicted next step-size, wrapped up in a data structure constructed by make-state-and-step. The result produced by the integration driver is a procedure next which, given an initial state and an initial step-size, evolves the sequence of states, passing each state to an output procedure (which, for example, produces graphical output). At each time-step, the integration is performed using the step-size predicted during the previous step, provided that this is less than the specified max-h.[5]

The workbench includes various methods of integration that can be combined with integration drivers. Some of these methods are themselves automatically generated by operating upon simple integrators with *integrator transformers.*

One integrator transformer incorporates a general strategy described in [6], for transforming a non-adaptive integrator into an integrator with adaptive step-size control: Given a step-size h, perform one integration step of size h and compare the result with that obtained by performing two steps of size $h/2$. If the difference between the two results is smaller than a prescribed error tolerance, then the integration step succeeds, and we can attempt to use a larger value of h for the next integration step. If the difference is larger than the error tolerance, we choose a smaller value of h and try the integration step again.

---

[5]System-integrator is only one of a number of possible integration drivers. The one actually used in the workbench produces a *stream* of states, so that integration steps are performed on a "demand-driven" basis. (See [1] for information on stream processing.)

More precisely, let **2halfsteps** be the (vector) result of taking two steps of size $h/2$, and let **fullstep** be the result of taking one step of size h. Then

$$E = \max_i \left| \frac{\texttt{2halfsteps}_i - \texttt{fullstep}_i}{\texttt{2halfsteps}_i} \right|$$

is an estimate of the relative error. Let $\texttt{err} = \frac{E}{\texttt{tolerance}}$ be the ratio of E to a prescribed error tolerance. We choose the new step-size to be

$$\texttt{newh} = h \times \texttt{err}^{\frac{-1}{n+1}} \times \texttt{safety}$$

where the underlying method of integration has order n, and where **safety** is a safety factor slightly smaller than 1. If the integration step fails ($\texttt{err} > 1$) we retry the step with **newh**. If the integration step succeeds ($\texttt{err} < 1$) we use **newh** for the next step. We can also make an order-$(n+1)$ correction to **2halfsteps**, computing the new state components as

$$\texttt{2halfsteps}_i + \frac{\texttt{2halfsteps}_i - \texttt{fullstep}_i}{2^n - 1}$$

See [6] for more details.

The **make-adaptive** procedure, which implements this strategy, is shown in Figure 5. The arguments to **make-adaptive** are a **stepper** that performs one step of a non-adaptive method of integration, together with the order of the method. **Make-adaptive** returns the corresponding adaptive integrator, which takes a system derivative f as argument and returns a procedure which, given a state and stepsize, returns the next state and a new stepsize.[6]

The **stepper** to be transformed by **make-adaptive** is a procedure that takes as arguments a system derivative f, a state y, a stepsize h, and the value dy/dt of f at y.[7] Here is a simple first-order backward

---

[6]Some details of **make-adaptive**: **Zero-stop** is a small number that is used to avoid possible division by zero. **Scale-vector** is a procedure which, given a number, returns a procedure that scales vectors by that number. **Elementwise** takes as argument a procedure of n arguments. It returns the procedure of n vectors that applies the original procedure to the corresponding elements of the vectors and produces the vector of results.

[7]One could easily arrange for the stepper itself to compute dy/dt. The reason for passing dy/dt as an argument is to avoid computing it twice in each adaptive integration step—once when evaluating **fullstep** and once when evaluating **halfstep**.

```
(define (make-adaptive stepper order)
  (let ((error-scale (/ -1 (+ order 1)))
        (scale-diff (scale-vector (/ 1 (- (expt 2 order) 1))))))
    (lambda (f)
      (lambda (state h-init)
        (let ((der-state (f state)))
          (let reduce-h-loop ((h h-init))
            (let* ((h/2 (/ h 2))
                   (fullstep (stepper f der-state state h))
                   (halfstep (stepper f der-state state h/2))
                   (2halfsteps (stepper f (f halfstep) halfstep h/2))
                   (diff (sub-vectors 2halfsteps fullstep))
                   (err (/ (maxnorm
                             ((elementwise (lambda (y d)
                                             (/ d (+ zero-stop (abs y)))))
                               2halfsteps
                               diff))
                            tolerance))
                   (newh (* safety h (expt err error-scale))))
              (if (> err 1)
                  (reduce-h-loop newh)
                  (make-state-and-step
                   (add-vectors 2halfsteps (scale-diff diff))
                   newh)))))))))
```

Figure 5. This is an integrator transformer procedure, which transforms a
non-adaptive integration **stepper** into an integration method with adaptive
step-size control.

Euler predictor-corrector stepper. Given a y and f, the stepper first computes a predicted next state $y_p = y + hf(y)$ and then estimates a corrected next state as $y + hf(y_p)$.

```
(define (backward-euler f dy/dt y h)
  (let* ((h* (scale-vector h))
         (yp (add-vectors y (h* dy/dt))))
    (add-vectors y (h* (f yp)))))
```

The corresponding adaptive integrator is constructed by

```
(define adaptive-backward-euler (make-adaptive backward-euler 1))
```

Here is a fourth-order Runge-Kutta stepper

```
(define 2* (scale-vector 2))
(define 1/2* (scale-vector (/ 1 2)))
(define 1/6* (scale-vector (/ 1 6)))

(define (runge-kutta-4 f dy/dt y h)
  (let* ((h* (scale-vector h))
         (k0 (h* dy/dt))
         (k1 (h* (f (add-vectors y (1/2* k0)))))
         (k2 (h* (f (add-vectors y (1/2* k1)))))
         (k3 (h* (f (add-vectors y k2)))))
    (add-vectors y
                 (1/6* (add-vectors (add-vectors k0 (2* k1))
                                    (add-vectors (2* k2) k3))))))
```

The corresponding adaptive integrator is

```
(define adaptive-runge-kutta-4 (make-adaptive runge-kutta-4 4))
```

Other transformation strategies lead to other sophisticated integrators. For example, the Bulirsch-Stoer integrator can be constructed by transforming a simple modified-midpoint stepper by means of a Richardson extrapolation generator [6].

2.4   Generating Iteration Schemes

In a nonlinear system, one can rarely solve algebraically for the state-variable derivatives as elementary functions of the state variables to produce an explicit system derivative of the form $\mathbf{x}' = \mathbf{F}(\mathbf{x})$. Instead, one encounters a system of nonlinear equations $\mathbf{E}(\mathbf{x}, \mathbf{x}') = \mathbf{0}$ where $\mathbf{x}$ is the vector of state variables, $\mathbf{x}'$ is the vector of corresponding derivatives, and $\mathbf{E}$ is a vector-valued function (one component for each scalar equation). Such systems of implicit differential equations can be attacked with iterative numerical schemes. In the Newton-Raphson scheme, for instance, one solves a system of nonlinear equations $\mathbf{G}(\mathbf{z}) = \mathbf{0}$ by choosing an initial guess $\mathbf{z}^{(0)}$ for the solution and iterating to approximate a fixed point of the transformation

$$\mathbf{z} \leftarrow \mathbf{z} - [\mathbf{DG}(\mathbf{z})]^{-1}\mathbf{G}(\mathbf{z})$$

where $\mathbf{DG}$ is the Jacobian derivative (matrix) of $\mathbf{G}$. This process can be carried out purely numerically, but it is greatly advantageous to use symbolic algebra to develop an explicit expression for $\mathbf{DG}$ and its inverse, because this avoids the need, at each iteration step, to numerically approximate the derivatives $\partial G_i/\partial z_j$ comprising the components of $\mathbf{DG}$.

The workbench uses this mixed symbolic-numerical method. In general, when attempting to compute the system derivative as outlined in section 2.1 the equation solver will fail to eliminate all non-state variables from the equations, and be left with a system of the form

$$\begin{aligned} \mathbf{G}(\mathbf{x}, \mathbf{u}) &= \mathbf{0} \\ \mathbf{x}' &= \tilde{\mathbf{F}}(\mathbf{x}, \mathbf{u}). \end{aligned}$$

Here $\mathbf{x}$ is the vector of state variables and $\mathbf{u}$ is a vector of additional "unknown" variables that could not be eliminated, leaving the implicit equations $\mathbf{G} = \mathbf{0}$. (Those derivatives of state variables that could not be eliminated in terms of state variables are included in the unknowns.) The easy case, in which all non-state variables are eliminated, corresponds to $\mathbf{u}$ being null. The workbench uses symbolic differentiation to derive expressions for the components of $\mathbf{DG}$, which it uses in turn to derive a symbolic expression for the Newton-Raphson transformation

$$\mathbf{u} \leftarrow \mathbf{u} - [\mathbf{DG}(\mathbf{x}, \mathbf{u})]^{-1}\mathbf{G}(\mathbf{x}, \mathbf{u}).$$

The workbench also derives symbolic expressions $\mathbf{u}' = \mathbf{H}(\mathbf{x}, \mathbf{u})$ for the derivatives $\mathbf{u}'$. This is accomplished by differentiating the equation $\mathbf{G}(\mathbf{x}, \mathbf{u}) = \mathbf{0}$ to obtain

$$\mathbf{DG_x} \cdot \mathbf{x}' + \mathbf{DG_u} \cdot \mathbf{u}' = \mathbf{0}$$

solving this for $\mathbf{u}'$, and eliminating the $\mathbf{x}'$ in terms of $\mathbf{x}$ and $\mathbf{u}$.

The actual system derivative computation proceeds as follows: The "system state" to be evolved consists of the state variables $\mathbf{x}$ augmented by the variables $\mathbf{u}$. Given values for state variables and guesses $\mathbf{u}^{(0)}$ for the unknowns, Newton-Raphson iteration produces values $\mathbf{u}$ that satisfy $\mathbf{G}(\mathbf{x}, \mathbf{u}) = 0$. The equations $\mathbf{x}' = \tilde{\mathbf{F}}(\mathbf{x}, \mathbf{u})$ and $\mathbf{u}' = \mathbf{H}(\mathbf{x}, \mathbf{u})$ now provide the required $\mathbf{x}'$ and $\mathbf{u}'$. Observe that each integration step evolves not only an updated $\mathbf{x}$, but also an updated $\mathbf{u}$ to be used as the initial guess $\mathbf{u}^{(0)}$ to begin the Newton-Raphson iteration at the next time-step. Usually, the integrator itself will produce a sufficiently good value $\mathbf{u}^{(0)} \approx \mathbf{u}$ that the Newton-Raphson correction will be iterated only once, if at all, at any given time-step.

The workbench compiles a system derivative generator procedure that incorporates the symbolically derived expressions for $\tilde{\mathbf{F}}$, $\mathbf{H}$, and the Newton-Raphson transformation. The system-derivative generator takes the network parameters as arguments and returns a system-derivative procedure that takes an augmented state as argument and produces the derivatives of the augmented state variables. Packaging things this way provides an important modularity—to evolve the system dynamics, the workbench can pass the system derivative to any general-purpose integration routine. The same integrators are used with the explicit system derivatives generated as in Section 2.1 and with the implicit system derivatives that incorporate iterative schemes.

Example: A Circuit with Cube-Law Resistors

To illustrate the above strategy, consider the nonlinear RLC circuit shown in Figure 6, containing a voltage source, a capacitor, an inductor, and two nonlinear resistors. The resistors are each cube-law resistors with v-i characteristic $v.a = i.b + (i.b)^3$ where a and b are parameters that scale voltage and current:[8]

```
(define-network cubic-rlc
  ()
  (n1 n2)
```

---

[8] The primitive part employed here, `non-linear-resistor`, is a device with nodes n+ and n-. Its parameters are a voltage **v**, a current **i** and a v-i characteristic **vic**, which is a procedure applied to **v** and **i** to produce an algebraic constraint.

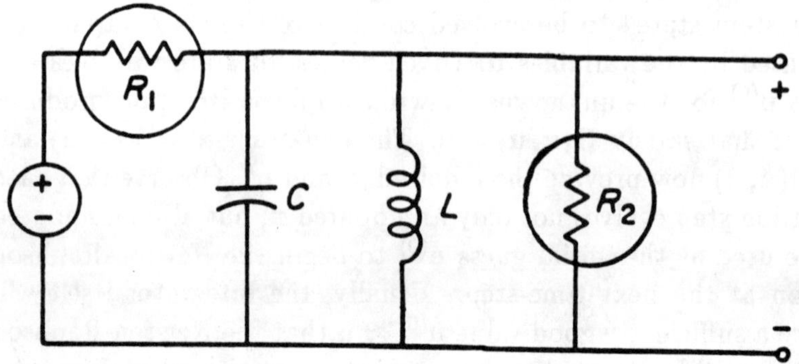

Figure 6. This second-order circuit contains two nonlinear resistors, each with a cubic *v-i* characteristic. Since the workbench's algebraic manipulator does not solve general cubic equations in closed form, the system derivative generated for this circuit incorporates a symbolically generated Newton-Raphson interation scheme.

```
(parts (s voltage-source (n+ n1) (n- gnd))
       (r1 cube-law-resistor (n+ n1) (n- n2))
       (c capacitor (n+ n2) (n- gnd))
       (l inductor (n+ n2) (n- gnd))
       (r2 cube-law-resistor (n+ n2) (n- gnd)))))
```

The equation solver attacks the resulting equations, in which the state variables are the inductor current i_L and the capacitor voltage v_C. The solver succeeds in eliminating di_L/dt, dv_C/dt, and all the non-state circuit variables except for two. These two "unknown" variables are the resistor current i_{R_2} and the source current i_S. The final two equations comprising the system $\mathbf{G}(\mathbf{x}, \mathbf{u}) = \mathbf{0}$ on which the solver cannot make further progress are

$$a_{R_2} b_{R_2}^2 i_{R_2} + a_{R_2} i_{R_2}^3 = b_{R_2}^3 v_C$$
$$a_{R_1} b_{R_1}^2 i_S + a_{R_1} i_S^3 + b_{R_1}^3 v_S = b_{R_1}^3 v_C.$$

Following the method outlined above, the workbench differentiates $\mathbf{G}$ with respect to $\mathbf{u} = (i_{R_2}, i_S)$ and with respect to $\mathbf{x} = (i_L, v_C)$ to produce the Newton-Raphson transformation and the derivatives $\mathbf{u}'$, and compiles the resulting expressions to form the system-derivative generator shown in Figure 7. This procedure takes the system parameters as arguments and returns a procedure that implements the update strategy: Given an augmented *state* vector $(t, i_L, v_C, i_{R_2}, i_S)$, extract from this the three state components (t, i_L, v_C) and the two *unknown* variables (i_{R_2}, i_S). The unknowns are used to initialize a vector-fixed-point operation whose returned *values* are the corrected unknowns—a fixed point of the Newton-Raphson transformation. The result returned by the system derivative is a vector whose components are the derivatives of the five variables in the augmented *state*, computed as a functions of the given t, i_L, v_C, and of the two corrected unknowns.[9]

3   Frequency-Domain Analysis

In addition to developing time-domain simulations, the workbench can perform frequency-domain analyses of linear systems. It does this by constructing an algebraic environment that contains the bilateral Laplace transforms of the constraint equations and algebraically solves these equations for the transforms of the circuit variables.

For instance, to analyze the Twin-T network, the workbench must deal with equations such as the transformed fact156 given above in Section 1:

```
(asserting fact287
          (= (transform ($ t s) (i.c3 t))
             (transform ($ t s)
                        (* c.c3 (rate (v.c3 t) ($ t))))))).
```

To handle these constraints, the workbench's algebraic manipulator performs such simplifications as

```
==> (algebra-value
      '(transform ($ t s)
                  (* c.c3 (rate (v.c3 t) ($ t)))))
```

```
(* c.c3 s (transform ($ t s) (v.c3 t))).
```

---

[9]The update-state! expression in the system-derivative procedure updates the augmented state to reflect the correction of u obtained by Newton-Raphson iteration. This updating has no effect on the computations described in this paper. In the actual workbench integration driver, where we evolve and store a stream of states, the updating ensures that any procedures that later examine the stream of states will see the corrected values.

```
(lambda (b.r2 a.r2 l.l c.c b.r1 a.r1 strength.s)
 (let ((g121 (* b.r1 b.r1)) (g118 (* b.r2 b.r2)))
  (let ((g119 (* b.r1 g121)) (g117 (* a.r2 c.c))
        (g120 (* a.r1 c.c)) (g116 (* b.r2 g118)))
   (lambda (*state*)
    (let ((t (vector-ref *state* 0))
          (i.l (vector-ref *state* 1))
          (v.c (vector-ref *state* 2)))
      (let ((g113 (* -1 i.l)))
       (let ((*values*
               (vector-fixed-point
                (lambda (*unknowns*)
                  (let ((i.r2 (vector-ref *unknowns* 0))
                        (i.s (vector-ref *unknowns* 1)))
                   (let ((g123 (* a.r1 i.s i.s))
                         (g122 (* a.r2 i.r2 i.r2)))
                    (vector
                      (/ (+ (* 2 g122 i.r2) (* g116 v.c))
                         (+ (* 3 g122) (* a.r2 g118)))
                      (/ (+ (* -1 g119 (strength.s t))
                            (* 2 g123 i.s)
                            (* g119 v.c))
                         (+ (* 3 g123) (* a.r1 g121)))))))
                (vector (vector-ref *state* 3) (vector-ref *state* 4)))))
        (update-state! *state* 3 *values*)
        (let ((i.r2 (vector-ref *values* 0))
              (i.s (vector-ref *values* 1)))
          (let ((g115 (* -1 i.r2)) (g114 (* -1 i.s)))
           (vector 1
                   (/ v.c l.l)
                   (/ (+ g113 g114 g115) c.c)
                   (/ (+ (* g113 g116) (* g114 g116) (* g115 g116))
                      (+ (* 3 g117 i.r2 i.r2) (* g117 g118)))
                   (/ (+ (* g113 g119) (* g114 g119) (* g115 g119))
                      (+ (* 3 g120 i.s i.s) (* g120 g121)))))))))))))
```

Figure 7. The system-derivative generator compiled for the cubic-rlc network
incorporates an automatically constucted Newton-Raphson iteration.

The simplification rules for transforms are expressed in a pattern-match and substitution language. The two rules

```
((transform ($ ?t:symbol? ?s:symbol?) (impulse ($ ?t) ?t0))
 (independent? t0 t)
 '(exp (* -1 ,t0 ,s)))

((transform ($ ?t:symbol? ?s:symbol?) (rate ?exp ($ ?t)))
 no-restrictions
 '(* ,s (transform ($ ,t ,s) ,exp)))
```

illustrate the kinds of transformations that can be specified. These rules encode the transform equations

$$\mathcal{L}\left[\delta(t - t_0)\right] = e^{-t_0 s} \; ;$$

i.e., the transform of a shifted impulse is an exponential, and

$$\mathcal{L}[dx/dt] = s\mathcal{L}[x] \; ;$$

i.e., time-differentiation transforms to multiplication by s. In general, a rule consists of a *pattern* to be matched, an additional predicate that the matching values must satisfy, and a *replacement* to be instantiated with the matching values if the match is successful. In each of the two rules above, the pattern stipulates that the expressions matching t and s must satisfy the symbol? predicate. The first rule also specifies that the impulse offset, t0, must be independent of t.

Because its simplifier incorporates a general pattern-match language, the workbench can readily be extended to deal with new operators and special functions.[10] The same language is used to implement the simplification rules that handle derivatives in time-domain analysis. Here, for instance, is the rule for differentiating quotients

$$\frac{d(x/y)}{dt} = \frac{y(dx/dt) - x(dy/dt)}{y^2}$$

```
((rate (/ ?x ?y) ($ ?t))
 no-restrictions
 '(/ (- (* ,y (rate ,x ($ ,t)))
        (* ,x (rate ,y ($ ,t))))
     (* ,y ,y)))
```

---

[10]This follows Macsyma [5], which provides a pattern matcher that allows users to extend the simplifier.

After solving the frequency-domain equations, the workbench can compute the voltage-transfer ratio of the network as the quotient of two degree-three polynomials in s:

```
==> (algebra-value
      '(/ (- (transform ($ t s) (e.n4 t))
             (transform ($ t s) (e.gnd t)))
          (transform ($ t s) (v.s t)))))

(/ (+ (* s s s r.r1 c.c3 r.r3 c.c2 r.r2 c.c1)
      (* s s r.r1 r.r3 c.c2 c.c1)
      (* s s r.r3 c.c2 r.r2 c.c1)
      (* s r.r3 c.c2)
      (* s r.r3 c.c1)
      1)
   (+ (* s s s r.r1 c.c3 r.r3 c.c2 r.r2 c.c1)
      (* s s r.r1 c.c3 r.r3 c.c2)
      (* s s r.r1 c.c3 r.r3 c.c1)
      (* s s r.r1 c.c3 c.c2 r.r2)
      (* s s r.r1 r.r3 c.c2 c.c1)
      (* s s r.r3 c.c2 r.r2 c.c1)
      (* s r.r1 c.c3)
      (* s r.r1 c.c2)
      (* s r.r3 c.c2)
      (* s r.r3 c.c1)
      (* s c.c2 r.r2)
      1))
```

Beginning with such a symbolic analysis, we can explore the effects of adding further constraints. For instance, the Twin-T circuit can be used as a notch filter, if we specialize the resistances and capacitances so that there is a zero in the transfer function at the chosen frequency. We can accomplish this by asserting extra constraints in the algebraic environment

```
(= c.c2 c.c1)
(= c.c3 (* 2 c.c1))
(= r.r2 r.r1)
(= r.r3 (/ r.r1 2))
```

and eliminating the variables c.c2, c.c3, r.r2, and r.r3. In this case, the voltage-transfer ratio reduces to the quotient of degree-two polynomials

$$H(s) = \frac{s^2 R_1^2 C_1^2 + 1}{s^2 R_1^2 C_1^2 + 4s R_1 C_1 + 1}$$

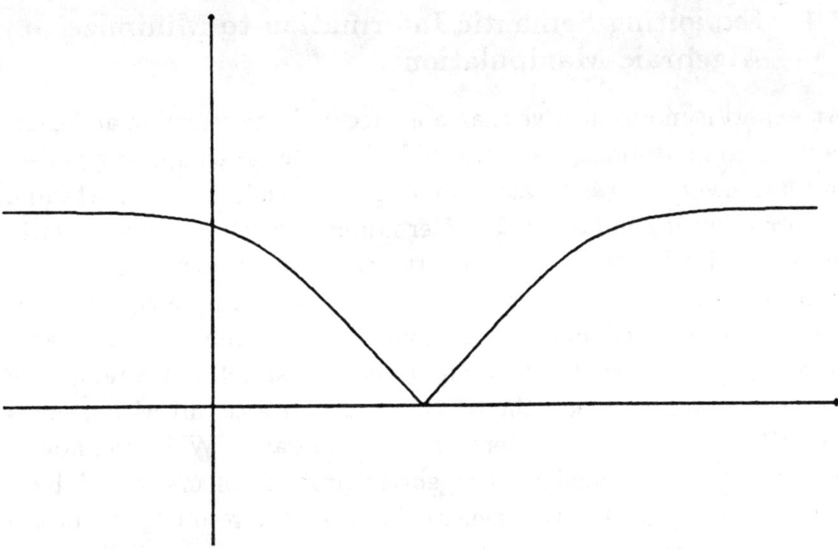

Figure 8. The frequency response of the Twin-T network graphed by the work-bench. The vertical axis is the magnitude of the voltage-transfer ratio. The vertical scale is $[-1, 2]$. The horizontal axis is the base-ten logarithm of the frequency in radians. The horizontal scale is $[-1, 3]$. The parameters are as in Figure 2. The notch formed by the zero at $\omega = 1/R_1 C_1$ explains the behavior of the output shown in Figure 2.

```
==> (algebra-value
      '(/ (- (transform ($ t s) (e.n4 t))
             (transform ($ t s) (e.gnd t)))
          (transform ($ t s) (v.s t))))

(/ (+ (* s s r.r1 r.r1 c.c1 c.c1) 1)
   (+ (* s s r.r1 r.r1 c.c1 c.c1)
      (* 4 s r.r1 c.c1)
      1))
```

As before, the workbench can use these expressions to compile pro-cedures that graph functions of frequency. Figure 8 shows a graph of the magnitude of $H(j\omega)$ versus $\log\omega$.

3.1 Exploiting Semantic Information to Minimize Algebraic Manipulation

An expert is more effective than a novice in doing scientific and engineering computations, not because he is better at computing per se, but because he knows what computing to do and, more importantly, what computing *not* to do. In determining the voltage-transfer ratio of an electrical network, a novice typically writes down many equations and attempts to solve them as a problem of pure algebra. For the expert electrical engineer, in contrast, the algebraic terms carry *meaning*. He knows, for example, that one cannot add a resistance to a capacitance, or that the transfer ratio for a circuit with a series capacitor has no constant term in the numerator. While the novice grapples with a complicated algebraic problem of many variables, the expert can postulate a general form for the result, and can use constraints and consistency checks to determine the detailed answer in a few steps.

Even for small networks, a fully symbolic frequency-domain analysis would exceed the capacity of all but the most powerful general-purpose algebraic manipulation systems. Dealing with rational functions of many variables is particularly troublesome in symbolic algebra, because, in order to avoid the explosion of intermediate expressions, one must repeatedly reduce quotients to lowest terms, which requires a multivariate greatest-common-divisor computation.[11]

Although the workbench performs symbolic algebra, it also exploits special properties of the domain to minimize the amount of raw algebraic manipulation required. For example, in the Twin-T circuit there are six symbolic device parameters and the frequency variable s. If the algebra is done without reducing rational functions using a full GCD algorithm, but rather by removing only the most obvious common factors, the expression for the voltage-transfer ratio turns out to be the ratio of two seventh-degree polynomials in s each with about 70 terms. This is obviously the wrong expression,

---

[11]Sussman and de Kleer [3] used the Macsyma symbolic computation system, running on a PDP10, to perform symbolic analysis and synthesis of electrical networks. For all but the very simplest networks, Macsyma was unable to perform the required reductions. Subsequently, Richard Zippel's *sparse modular algorithm* [7] enormously improved Macsyma's ability to compute multivariate GCDs. With current algorithms, a circuit of the complexity of the Twin-T network is near the limit of what Macsyma running on a PDP10 can cope with.

because there are only three capacitors, and so the degrees of the numerator and the denominator can be at most three in s. Moreover, by a theorem of P. M. Lin [4], each device parameter can occur to degree at most one.

Unfortunately, the degree requirements alone do not sufficiently constrain the algebra—for six device parameters, a polynomial of degree three in s can have up to 256 terms. To reduce the problem further, the workbench exploits constraints based on the dimensional information declared for each variable. For instance, the sum of a capacitance and a resistance cannot appear in a well-formed expression, because resistance and capacitance have different units; but the expression $RCs + 1$ is well-formed because the product of resistance and capacitance has the dimensions of time, and time is the inverse of frequency. The workbench's algebraic manipulator can determine the units of an algebraic expression. It computes the dimensions of the rational function to be reduced, thereby constraining the possible terms that can appear in the reduced form. In the case of the transfer ratio for the Twin-T network, the possible numerators and denominators turn out to have at most 20 terms each. Such small systems can be easily solved by numerical interpolation, even without a sophisticated GCD algorithm.

4   Periodic Orbits of Driven Oscillators

The elements that the workbench constructs for performing simulations can be incorporated into procedures that perform higher-level analyses of dynamical systems. In this section, we illustrate how the workbench automatically generates programs for investigating the behavior of periodically driven nonlinear oscillators. One way to study the dynamics of periodically driven oscillators is through the structures of periodic orbits whose periods are integer multiples of the drive period. Such orbits may be stable, in that small perturbations in initial conditions remain close to the periodic orbits, or they may be unstable. The workbench compiles procedures that find periodic orbits, determine their stability characteristics, and track how the orbits change as the parameters of system are varied.

Figure 9 shows the circuit diagram of a driven van der Pol oscillator, one of the simplest nonlinear systems that displays interesting behavior. The nonlinear resistor has a cubic v-i characteris-

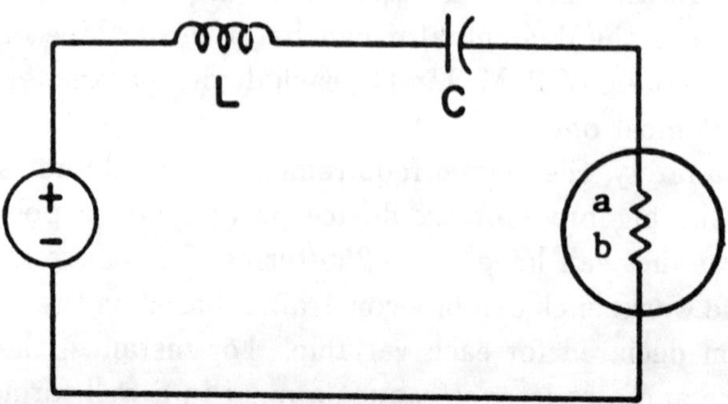

Figure 9. A driven van der Pol oscillator may be constructed from a series RLC circuit with a nonlinear resistor. Reprinted with permission from *Communications of the ACM*, Vol. 32, p. 558. © 1989 by the Association for Computing Machinery, Inc.

tic $v = ai^3 - bi$ that exhibits negative resistance for small currents and positive resistance for large currents. If there is no drive and the effective Q is large, the system oscillates stably at a frequency primarily determined by the inductance and the capacitance. In state space (v_C and i_L) the undriven oscillation approaches a stable limit cycle. Although the dynamics of the undriven system are well-understood, if we drive the system with a periodic drive, the competition between the drive and the autonomous oscillatory behavior leads to extremely complex, even chaotic behavior. In this section we will use the workbench to explore the behavior of the van der Pol oscillator when driven at a period close to a subharmonic of its autonomous oscillatory frequency.

4.1   Locating Periodic Orbits

One can find periodic orbits by a fixed-point search. Given an initial state $\mathbf{x}$, integrate the equations through one period of the drive and find the end state $\mathbf{S}(\mathbf{x})$. If the chosen initial state is a fixed point of this *period map* $\mathbf{S}$ then the orbit is periodic. Moreover, the stability of the periodic orbit can be determined by linearizing the period map in a neighborhood of this fixed point and examining the eigenvalues.[12]

---

[12]This is Floquet's method for analyzing nonlinear systems with periodic drives, generalized by Poincaré for other systems with periodic orbits.

Fixed-points can be found by Newton-Raphson iteration, provided we can compute the Jacobian derivative of the period map $\mathbf{x} \mapsto \mathbf{S}(\mathbf{x})$. This can be done by a mixture of numerical and symbolic computation. Since $\mathbf{S}$ is obtained by integrating the system derivative $\mathbf{x}' = \mathbf{F}(\mathbf{x})$, the Jacobian of $\mathbf{S}$ is obtained by integrating the associated *variational system*, which is the linear system

$$(\delta\mathbf{x})' = \mathbf{A} \cdot \delta\mathbf{x} \qquad \text{where } A_{ij} = \frac{\partial F_i}{\partial x_j}.$$

Thus we can compute the Jacobian matrix $\mathbf{D_x S}$ by integrating the variational system along the orbit, starting with an orthonormal basis. Even though the integration must be performed numerically, the variational system can be developed symbolically by differentiating the expressions in $\mathbf{F}$. The workbench prepares a system derivative augmented with a variational system for use in this fixed-point search.

We illustrate this strategy applied to the driven van der Pol system. Here is the system as described to the workbench:

```
(define-network driven-van-der-pol
   ((a parameter v/i^3) (b parameter resistance) (d drive voltage))
   (n1 n2 n3)
   (parts (nl-res nonlinear-resistor (n+ n3) (n- gnd)
               (vic (lambda (v i)
                      '(= ,v (- (* ,a ,i ,i ,i) (* ,b ,i)))))))
          (l inductor (n+ n2) (n- n3))
          (c capacitor (n+ n1) (n- n2))
          (s voltage-source (n+ n1) (n- gnd) (strength d))))
```

and here are the resulting expressions for the system derivative, as computed by the workbench:

```
(rate (v.c t) ($ t)) = (/ i.l c.c)
(rate (i.l t) ($ t)) = (/ (+ (* -1 a i.l i.l i.l)
                            (* b i.l)
                            (d t)
                            (* -1 v.c))
                          l.l)).
```

Just as with the Twin-T network of Section 2, we can use the system derivative to evolve the time-domain trajectories. Figure 10 shows a particular trajectory. We see from the figure that the trajectory approaches a periodic orbit.

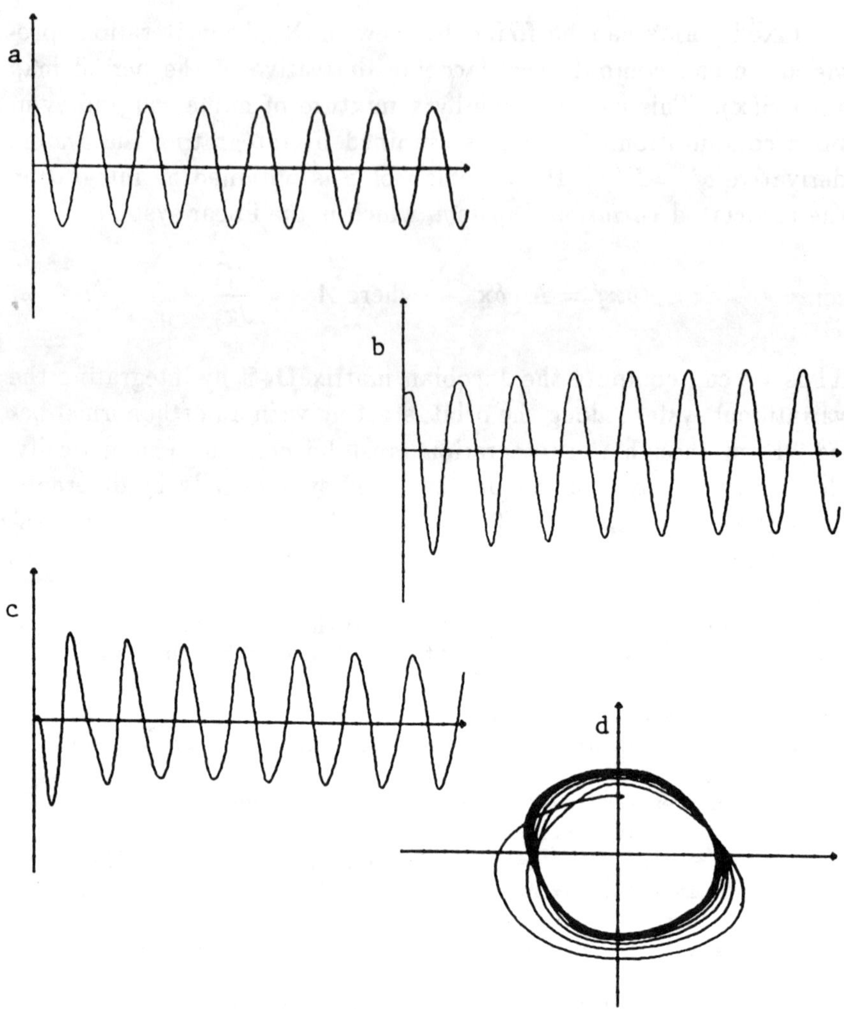

Figure 10. Time-domain plots of the driven van der Pol oscillator show the approach to a periodic orbit. Trace (a) shows the drive. Trace (b) shows the voltage across the capacitor. Trace (c) shows the current through the inductor. Trace (d) shows the state-space trajectory. The abscissa is the current through the inductor and the ordinate is the voltage across the capacitor. We show 30 seconds of simulated time. The voltage scales are $[-100, 100]$. The current scale is $[-0.2, 0.2]$ amperes. The parameters are $C = .001, L = 100, a = 10000, b = 100$. The drive is $d(t) = 40 \cos 1.6t$. In this example, we chose the drive frequency to be slightly higher than half the resonant frequency $1/\sqrt{LC}$. The initial state is $v_C = 37$ volts and $i_L = 0$ amps.

To search for a periodic orbit, the workbench compiles an augmented system derivative generator, shown in Figure 11, that (as a function of the parameters) computes a system derivative for the variational system. This is a procedure that given a *variational system state*

$$\texttt{*varstate*} = (t, v_C, i_L, v_C[\delta v_C], v_C[\delta i_L], i_L[\delta v_C], i_L[\delta i_L])$$

computes the state derivative.[13] Combining the system derivative with an integrator produces an end-point procedure that maps variational states to variational states by integrating over a given period. The result is passed through a Newton-Raphson transformation to realize the fixed-point search.

Starting the fixed-point search at the initial state $(37, 0)$ produces the following periodic point:

```
((periodic-point (54.0623 -2.40923e-3))
 (orbit-type spiral attractor)
 (eigenvalues ((*rect* .762164 .187228) mag .784824)
             ((*rect* .762164 -.187228) mag .784824))
 (trace 1.52433)
 (det .615949))
```

The Jacobian matrix not only directs the Newton-Raphson search for the fixed point, but it also provides information about the stability of the fixed point and the associated periodic orbit. For this fixed point, there are complex-conjugate eigenvalues with magnitude less than one. This is therefore a stable fixed point, indicating that the associated periodic orbit is a spiral attractor.[14]

4.2 Tracking Periodic Orbits

The workbench can also compile procedures to track how periodic orbits move as system parameters are varied.

Let $\mathbf{x} \mapsto S(\mathbf{p}, \mathbf{x})$ be the period map with explicit dependence on the parameters $\mathbf{p}$ of the system. Let $\mathbf{x}$ be a state such that $S(\mathbf{p}, \mathbf{x}) = \mathbf{x}$ for a particular choice of the parameters $\mathbf{p}$. For an incremental change $\Delta\mathbf{p}$ we compute, to first order, the corresponding incremental change $\Delta\mathbf{x}$ such that $\mathbf{x} + \Delta\mathbf{x} = S(\mathbf{p} + \Delta\mathbf{p}, \mathbf{x} + \Delta\mathbf{x})$.

---

[13]The meaning of the final four components of the state is that $v_C[\delta i_L]$, for example, is the component in the direction v_C of the variation vector δi_L.

[14]See Abraham and Shaw [2] for fewer details.

```
(lambda (c.c l.l d b a)
  (lambda (*varstate*)
    (let ((t (vector-ref *varstate* 0))
          (v.c (vector-ref *varstate* 1))
          (i.l (vector-ref *varstate* 2))
          (v.c.del.v.c (vector-ref *varstate* 3))
          (v.c.del.i.l (vector-ref *varstate* 4))
          (i.l.del.v.c (vector-ref *varstate* 5))
          (i.l.del.i.l (vector-ref *varstate* 6)))
      (let ((g27 (* a i.l i.l)))
        (let ((g28 (* -3 g27)))
          (vector 1
                  (/ i.l c.c)
                  (/ (+ (* -1 g27 i.l)
                        (* -1 v.c)
                        (* b i.l)
                        (d t))
                     l.l)
                  (/ v.c.del.i.l c.c)
                  (/ (+ (* -1 v.c.del.v.c)
                        (* b v.c.del.i.l)
                        (* g28 v.c.del.i.l))
                     l.l)
                  (/ i.l.del.i.l c.c)
                  (/ (+ (* -1 i.l.del.v.c)
                        (* b i.l.del.i.l)
                        (* g28 i.l.del.i.l))
                     l.l)))))))
```

Figure 11. This is the augmented system derivative generator compiled to
evolve variational states for the driven van der Pol oscillator. Reprinted with
permission from *Communications of the ACM*, Vol. 32, p. 559. © 1989 by the
Association for Computing Machinery, Inc.

$$\begin{aligned}
\mathbf{x} + \Delta\mathbf{x} &= \mathbf{S}(\mathbf{p} + \Delta\mathbf{p}, \mathbf{x} + \Delta\mathbf{x}) \\
&= \mathbf{S}(\mathbf{p},\mathbf{x}) + \mathbf{D_x S} \cdot \Delta\mathbf{x} + \mathbf{D_p S} \cdot \Delta\mathbf{p} \\
&= \mathbf{x} + \mathbf{D_x S} \cdot \Delta\mathbf{x} + \mathbf{D_p S} \cdot \Delta\mathbf{p}
\end{aligned}$$

where $\mathbf{D_x S}$ and $\mathbf{D_p S}$ are the matrices of partial derivatives of $\mathbf{S}$ with respect to $\mathbf{x}$ and $\mathbf{p}$. Subtracting $\mathbf{x}$ from both sides yields

$$\Delta\mathbf{x} = \mathbf{D_x S} \cdot \Delta\mathbf{x} + \mathbf{D_p S} \cdot \Delta\mathbf{p}$$

and we conclude that

$$\Delta\mathbf{x} = (1 - \mathbf{D_x S})^{-1}\mathbf{D_p S} \cdot \Delta\mathbf{p}.$$

We use the first-order approximation $\mathbf{x} + \Delta\mathbf{x}$ as the initial guess for the actual fixed-point, and iteratively correct the guess with Newton-Raphson.

The matrix $\mathbf{D_x S}$ is computed as in Section 4.1, by numerical integration of the symbolically derived variational equations. $\mathbf{D_p S}$ is also computed by a mixture of symbolic differentiation and numerical integration: $\mathbf{S}(\mathbf{p},\mathbf{x})$ is the integral of the parameterized system derivative $\mathbf{x}' = \mathbf{F}(\mathbf{p},\mathbf{x})$. Thus $\mathbf{D_p S}$ can be found by numerically integrating the symbolically obtained partial derivatives $\partial F_i/\partial p_j$ along the orbit.

For the driven van der Pol system there are four parameters—the capacitance C, the inductance L, and the resistor-characteristic parameters b and a. The machine-generated procedure shown in Figure 12 computes the system derivative, augmented by the $\partial S_i/\partial p_j$. The components of the augmented state, ***varstate***, are t, v_C, i_L, dv_C/dC, di_L/dC, dv_C/dL, di_L/dL, dv_C/db, di_L/db, dv_C/da, and di_L/da.

We can use this procedure to track a fixed point of the driven van der Pol oscillator as we decrease the capacitance so that the resonant frequency of the oscillator passes through a resonance with the second harmonic of the drive. We start at $C = .001$ and decrease it in steps of 5×10^{-6}.

```
((periodic-point (54.0623 -2.40923e-3))
 (orbit-type spiral attractor)
 (eigenvalues ((*rect* .762164 .187228) mag .784824)
```

```
                  ((*rect* .762164 -.187228) mag .784824))
 (trace 1.52433)
 (det .615949)
 (parameters .001 100 100 10000))
```

The next value of capacitance is .000995, which leads to an estimate for the new fixed-point

```
(estimating-next-point (53.9397 -2.53539e-3)).
```

Starting the Newton-Raphson iteration with this guess produces the actual fixed point:

```
((periodic-point (53.9402 -2.53249e-3))
 (orbit-type spiral attractor)
 (eigenvalues ((*rect* .793545 .160865) mag .809686)
             ((*rect* .793545 -.160865) mag .809686))
 (trace 1.58709)
 (det .655591)
 (parameters .000995 100 100 10000)).
```

The magnitude of the eigenvalues has increased and the magnitude of the phase angle has decreased so the rate of local contraction to the orbit, and the rotation rate, have both slowed.

As we decrease the capacitance, this trend continues until the eigenvalues become real—the local rotation stops and the topological type of the orbit changes to a node, first a barely stable node and then a saddle:

```
((periodic-point (53.4632 -2.97031e-3))
 (orbit-type nodal attractor)
 (eigenvalues (.992466 mag .992466)
             (.841266 mag .841266))
 (trace 1.83373)
 (det .834928)
 (parameters .000975 100 100 10000))

((periodic-point (53.3453 -3.06418e-3))
 (orbit-type nodal saddle)
 (eigenvalues (1.05071 mag 1.05071)
             (.84258 mag .84258))
 (trace 1.89329)
 (det .885305)
 (parameters .00097 100 100 10000))
```

```scheme
(lambda (c.c l.l d b a)
  (lambda (*varstate*)
    (let ((t (vector-ref *varstate* 0))
          (v.c (vector-ref *varstate* 1))
          (i.l (vector-ref *varstate* 2))
          (v.c.c.c (vector-ref *varstate* 3))
          (i.l.c.c (vector-ref *varstate* 4))
          (v.c.l.l (vector-ref *varstate* 5))
          (i.l.l.l (vector-ref *varstate* 6))
          (v.c.b (vector-ref *varstate* 7))
          (i.l.b (vector-ref *varstate* 8))
          (v.c.a (vector-ref *varstate* 9))
          (i.l.a (vector-ref *varstate* 10)))
      (let ((g34 (* i.l i.l)))
        (let ((g30 (* a g34)))
          (let ((g32 (d t)) (g29 (* -1 i.l))
                (g33 (* i.l.l.l l.l)) (g31 (* -3 g30)))
            (vector 1
                    (/ i.l c.c)
                    (/ (+ g32 (* -1 v.c) (* b i.l) (* g29 g30)) l.l)
                    (/ (+ g29 (* c.c i.l.c.c))
                       (* c.c c.c))
                    (/ (+ (* -1 v.c.c.c) (* b i.l.c.c) (* g31 i.l.c.c))
                       l.l)
                    (/ i.l.l.l c.c)
                    (/ (+ v.c
                          (* -1 g32)
                          (* -1 l.l v.c.l.l)
                          (* b g29)
                          (* b g33)
                          (* g30 i.l)
                          (* g31 g33))
                       (* l.l l.l))
                    (/ i.l.b c.c)
                    (/ (+ i.l (* -1 v.c.b) (* b i.l.b) (* g31 i.l.b)) l.l)
                    (/ i.l.a c.c)
                    (/ (+ (* -1 v.c.a) (* b i.l.a)
                          (* g29 g34) (* g31 i.l.a))
                       l.l)))))))))
```

Figure 12. This is the augmented system derivative generator compiled to
track how periodic points of the driven van der Pol system vary with the system
parameters.

Observe that this transition happens just as the resonant frequency $1/\sqrt{LC}$ passes through twice the drive frequency.

As we further decrease the capacitance, the eigenvalues increase until they are both greater than one; the topological type changes again to a nodal, then a spiral repellor:

```
((periodic-point (52.7759 -3.47088e-3))
 (orbit-type nodal repellor)
 (eigenvalues (1.08793 mag 1.08793)
              (1.07915 mag 1.07915))
 (trace 2.16708)
 (det 1.17404)
 (parameters .000945 100 100 10000))

((periodic-point (52.6649 -3.53871e-3))
 (orbit-type spiral repellor)
 (eigenvalues ((*rect* 1.10764 .113169) mag 1.1134)
              ((*rect* 1.10764 -.113169) mag 1.1134))
 (trace 2.21527)
 (det 1.23967)
 (parameters .00094 100 100 10000))
```

Further decreasing the capacitance leads to even larger eigenvalues with higher rotation rates.

Acknowledgments

The dynamicist's workbench is part of a larger project at M.I.T. to investigate the use of combined numerical and symbolic methods in scientific and engineering computing. Tim Chinowsky, David Espinosa, Bill Siebert, Matthew Halfant, Jacob Katzenelson, and Jack Wisdom contributed to the development of the workbench software. We also thank Bill Dally, Yekta Gürsel, Ken Yip, Eric Grimson, Julie Sussman, Josh Barnes, and Piet Hut for their help with the presentation.

References

[1] Abelson, Harold and Sussman, Gerald Jay with Sussman, Julie, *Structure and Interpretation of Computer Programs*, MIT Press, Cambridge, Mass., 1985.

[2] Abraham, Ralph H. and Shaw, Christopher D., *Dynamics—The Geometry of Behavior. Part II: Chaotic Behavior.* Aerial Press, Santa Cruz, Cal., 1983.

[3] de Kleer, Johan and Sussman, Gerald Jay, "Propagation of constraints applied to circuit synthesis," *Circuit Theory and Applications*, Vol. 8, pp. 127–144, 1980.

[4] Lin, P. M., "A survey of applications of symbolic network functions." *IEEE Transactions on Circuit Theory*, Vol. CT-20, No. 6, pp. 732–737, November, 1973.

[5] Mathlab Group, *Macsyma Reference Manual*, Laboratory for Computer Science, Massachusetts Institute of Technology, 1977.

[6] Press, William H., Flannery, Brian P., Teukolsky, Saul A., and Vetterling, William T., *Numerical Recipes: The Art of Scientific Computing*, Cambridge University Press, 1985.

[7] Zippel, Richard, *Probabilistic Algorithms for Sparse Polynomials*, Ph.D. thesis, Massachusetts Institute of Technology, 1979.

Chapter 3

Symbolic Computations in Differential Geometry Applied to Nonlinear Control Systems

O. Akhrif and G.L. Blankenship

1 Introduction

There are now several programs and CAD packages integrating aspects of numerical mathematics and control system design:

- ORACLS by NASA.

- DELIGHT.MIMO by the University of California, Berkeley and Imperial College, London.

- DELIGHT.MaryLin by the University of Maryland.

- CACE-III and MEAD by the General Electric Company, Northrop, and RPI.

- and various commercial packages like Matrix$_X$, Program CC, and Cntrl-C.

Perhaps less well known is the recent progress of the application of symbolic calculations in mathematical analysis, differential equations, and particularly differential geometry and its applications in nonlinear control theory. In this paper, we shall describe CONDENS, a software system written in the computer algebra system Macsyma for the analysis and design of nonlinear control systems based on differential geometric methods.

Differential geometry provides a very useful framework for the analysis of nonlinear control systems. It permits the generalization of many constructions for linear control systems to the nonlinear case. However, differential geometric objects (Lie brackets, Lie derivatives, etc.) are not easily manipulated by hand. CONDENS employs Macsyma as a powerful tool for manipulating differential geometric computations. CONDENS uses differential geometric tools expressed as function blocks in Macsyma for the analysis and design of nonlinear control systems, emphasizing the design of controllers for the output tracking problem.

The programs that form CONDENS fall into three different categories:

First, we have a set of user-defined functions that perform certain differential geometric computations, including straightforward computations such as *Lie brackets* and *Lie derivatives* and more complex computations such as *Kronecker indices* or *relative order* of nonlinear systems.

Second, CONDENS contains more sophisticated modules that use the basic functions to address two theoretical issues in geometric control theory — *feedback equivalence* and *left-invertibility* — for nonlinear systems affine in control, i.e., systems described (in local coordinates) by:

$$\frac{dx}{dt} = f(x) + \sum_{i=1}^{m} u_i(t)g_i(x). \tag{1}$$

Two design packages compose the third category of programs in CONDENS. Their purpose is to use either feedback linearization or invertibility of nonlinear systems to design a control law that forces the output of a nonlinear system to follow some desired trajectory.

These design packages include the capability to automatically generate Fortran programs for the solution of problems posed in symbolic form or for simulation purposes. Thus, for a new design problem, the time involved in analyzing the analytical structure of the problem and then writing the Fortran code to execute the design algorithm is eliminated. The time required to write, test, and debug the associated Fortran code is eliminated.

The advantages of CONDENS are clear. It can make available to the design engineer design methods that rely on more or less sophisticated mathematical tools. Most importantly, the system allows the engineer to interact with the computer for design at the symbolic manipulation level. In this way, he can modify his analysis or design problem by modifying the symbolic functional form of the model. The Fortran subroutines that he might have to modify by hand to accomplish this in conventional design procedures are written automatically for him.

There are several computer algebra systems, e.g., Formac, Macsyma, Reduce, which might have been used in this work. Macsyma appears to be the system of choice in several centers of control system work. It can implemented on computer systems supporting Lisp, it is readily available and, consequently, widespread. Last, but not least, we are charmed by the interactive facilities of Macsyma.

In the next section, we recall briefly the concepts of state and feedback equivalence among control systems and right- and left-invertibility of nonlinear systems. We also show how we can use these concepts to design an output tracking controller for the non-linear system (1).

In Section 3, we present a summary of the programming work. A description of the different modules of the program is given together with examples and simulation results to illustrate the performance of the system.

2   Theory

2.1   Feedback Equivalence Among Nonlinear Systems

Feedback equivalence is an equivalence relation (transitive, symmetric, and reflexive) among control systems which generalizes the concept of the linear feedback group in linear system theory [12] leading, among other things, to the Brunovsky canonical form and the definition of controllability indices. Several authors ([13], [2], [6], [4], [5], and [10]) introduced the concept of feedback linearization or feedback equivalence for nonlinear systems and gave necessary and sufficient conditions determining those nonlinear systems which are feedback equivalent to linear controllable ones.

The nonlinear system affine in control:

$$\dot{x} = f(x) + \sum_{i=1}^{m} g_i(x)u_i \qquad x \in R^n \qquad (2)$$

is said to be *feedback linearizable* in a neighborhood U of the origin ($f(0) = 0$) if it can be transformed via change of coordinates in the state space, change of coordinates in the control space, and feedback to a controllable linear system in Brunovsky canonical form:

$$\dot{z} = Az + Bv \qquad z \in R^n \tag{3}$$

with *Kronecker* indices $k_1, \ldots, k_m$.

Hunt, Sun, and Meyer in [4] and [5] gave necessary and sufficient conditions on the Lie structure of system (2) under which a transformation of the type we consider exists. They also gave a procedure for construction of such a transformation. This involves solving a set of partial differential equations, or equivalently n sets of n ordinary differential equations. This is not always possible or easy to do. However, in some cases—the single input case ([2]) or systems in block triangular form ([8] and [9])—the construction of the transformation is simple.

CONDENS contains a module (*TRANSFORM*) that treats the feedback linearization problem. It first investigates the existence of the linearizing one to one transformation by checking the necessary and sufficient conditions. If the transformation is possible, it attempts to construct the transformation (change of coordinates and nonlinear feedback) and its inverse.

Application: Design of Output Tracking Controllers

The design technique is to build a controller for the nonlinear system in terms of one for the equivalent linear canonical system and then inverting the transformation. The design proceeds in three steps. First, the given nonlinear system is transformed into a constant, decoupled, controllable linear representation. Second, standard linear design techniques, such as pole placement, are used to design an output tracking control for the equivalent linear system, forcing its output to follow a desired trajectory. Third, the resulting control law is transformed back out into the original coordinates to obtain the control law in terms of the original controls.

G. Meyer at the NASA Ames Research Center proposed this scheme in the design of exact model following automatic pilots for vertical and short take-off aircraft and has applied it to several aircraft of increasing complexity ([8] and [9]).

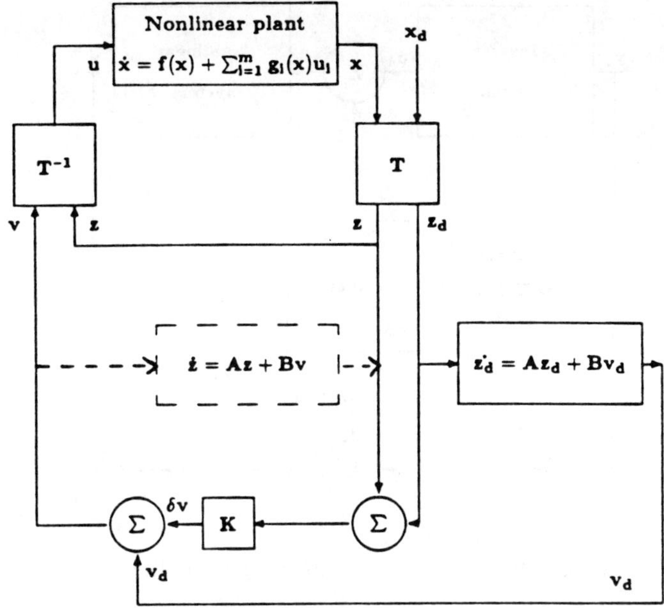

Figure 1. Design scheme used by FENOLS.

This design procedure is performed by FENOLS, one of the design packages contained in CONDENS (see Figure 1). A description of FENOLS and examples of its functions are given in Section 3.

2.2 Invertibility of Nonlinear Systems

In this subsection, we briefly review some of the results on left and right invertibility for nonlinear systems that are also implemented in CONDENS.

We consider systems of the following form:

$$\dot{x}(t) = f(x(t)) + \sum_{i=1}^{m} g_i(x(t))u_i(t) \qquad x(0) = x_0 \in M$$

$$y(t) = h(x(t)) \qquad u \in R^m, y \in R^m \tag{4}$$

where the state space M is a connected real analytic manifold, f, g are analytic vector fields on M, $h_i, i = 1, \ldots, m$ is a real analytic mapping, and $u \in U$, the class of real analytic functions from $[0, \infty)$ into R^m.

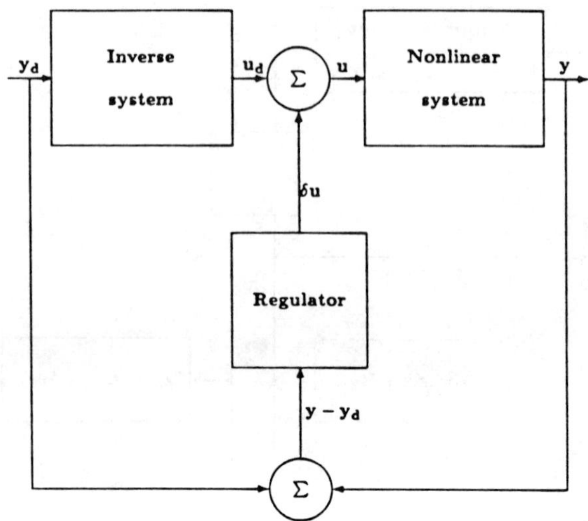

Figure 2. Design scheme using TRACKS.

The left-invertibility problem is the problem of injectivity of the
input-output map of system (4). If the input-output map is invertible
from the left, then it is possible to reconstruct uniquely the input
acting on the system from knowledge of the corresponding output.
The main application of this is in the output tracking problem, where
we try to control a system so that its output follows some desired
path. The inverse system is used to generate the required control
given the desired output function.

The problem of right-invertibility is the problem of determining functions $f(t)$ which can be realized as output of the nonlinear system (4) driven by a suitable input function. While the left-
invertibility is related to the *injectivity* of Input/Output map of the
nonlinear system (4), the right-invertibility is related to the *surjec-
tivity* of the Input/Output map.

Hirschorn [3] gave necessary and sufficient conditions of left- and
right-invertibility of system (4), as well as a procedure to construct
the left-inverse system. The notion of *relative degree* has a very
important role in this construction. For the multivariable case, refer
to ([13] and [14]).

CONDENS contains a module, INVERT, that checks if the nonlinear system (4) is left- or right-invertible and constructs the left

inverse system.

Application: Design of Output Tracking Controllers

It is easy to see how the notion of left- and right-invertibility can be applied to the output tracking problem. The application of the inversion algorithm to the nonlinear system (see ([13] and [14]) for the algorithm in the multivariable case) gives rise to a left-inverse system which, when driven by appropriate derivatives of the output y, produces $u(\cdot)$ as its output. The question of trajectory following by the output is related to right-invertibility of the nonlinear input-output map, and the ability of the nonlinear system to reproduce the reference path as its output. To obtain robustness in the control system under perturbations, design of a servocompensator around the inner loop using servomechanism theory is suggested.

TRACKS, the second design package contained in CONDENS, uses this procedure along with the module INVERT to design an output tracking controller (see Figure 2).

3 CONDENS: A Software Package Using Symbolic Manipulations

In this section CONDENS, a software package based on computer algebra, is presented. Employing the analytical methods presented above, CONDENS can, given a nonlinear system affine in control, answer questions such as: Is this system feedback-equivalent to a controllable linear one? If so, can we construct the diffeomorphism that makes this equivalence explicit? Is the nonlinear system invertible? What is the relative order of the system? In case the system is invertible, can we construct a left-inverse for it? Given a real analytic function $y(t)$, can $y(t)$ appear as output of the nonlinear system? If so, what is the required control?

CONDENS tries to answer these questions and uses this knowledge to solve a design problem: the output tracking problem for the given nonlinear control system. If we are interested in simulation results, the system can automatically generate Fortran programs for that purpose.

Before describing CONDENS in more detail, we shall make some general remarks relevant to all worked examples in each section.

1. Input lines always begin with a "(C_n)," which is the prompt character of Macsyma, indicating that the system is waiting for a command.

2. Results of commands terminated by a ";" are printed.

3. Results of commands terminated by a "$" are not printed.

4. If we are working with a single-input, single-output system:

$$\dot{x} = f(x) + g(x)u$$
$$y = h(x)$$

then the vector fields f and g are entered in the form of lists:

$$f : [f_1(x), f_2(x), \ldots, f_n(x)];$$
$$g : [g_1(x), g_2(x), \ldots, g_n(x)];$$
$$h : h(x);$$

5. If we are working with a multivariable control system:

$$\dot{x} = f(x) + \sum_{i=1}^{m} g_i(x)u_i$$
$$y = h(x)$$

$x \in R^n, \ u \in R^m, \ y \in R^m$

then the vecter field f is entered in the form:

$$f : [f_1(x), \ldots, f_n(x)];$$

but the m vector fields $g_1, g_2, \ldots, g_m$ are entered in the compact form:

$$g : [g_1, \ldots, g_m]; \quad \text{where each } g_i \text{ is} \quad g_i : [g_{i1}, \ldots, g_{in}];$$

3.1 Package Facilities

CONDENS is formed by a set of large functions, which, in turn are built up of smaller sub-functions. In order to save time, we would like, before using a certain function, to be able to automatically load into Macsyma only the subfunctions that this specific function uses.

This way, we will not have to load the entire package into Macsyma each time we want to use it.

For this, the **setup-autoload** Macsyma command is used. All the functions in CONDENS are collected in a special file

$$(\texttt{``initfile.mac''})$$

with file addresses for all the functions. Once this file is loaded into Macsyma, a CONDENS function, not previously defined, will be automatically loaded into Macsyma when it is called. So the first thing the user has to do once in Macsyma, is to load the login file (`load ''initfile.mac'';`). He can then start the package using the command `''condens();''` which will give him some hints on how to use CONDENS.

The file `''initfile.mac''` is listed below:

```
/*-*-*-*-*-*-*-*-*-*-*-*-*-*-*-*-*-*-*-*-*-*-*-*-*-*-*-*-*-*-*-*-*-*-*-*-*
 This is an init file for the package CONDENS:CONTROL DESIGN OF NONLINEAR
SYSTEMS. This is the only file that has to be loaded in Macsyma by the
user, all the other files will be automatically loaded as they are needed.
 So to get started, the user just has to type once : LOAD("initfile.mac"),
followed by the command CONDENS() which will give him some hints on how
to use the package.
 -*-*-*-*-*-*-*-*-*-*-*-*-*-*-*-*-*-*-*-*-*-*-*-*-*-*-*-*-*-*-*-*-*-*-*-*

loadprint:false$
setup_autoload("CONDENS/menu.mac",menu)$
setup_autoload("CONDENS/geofun1.mac",jacob,pjacob,lie,plie,adj)$
setup_autoload("CONDENS/geofun2.mac",lidev,plidev,nlidev,relord)$
setup_autoload("CONDENS/transform.mac",transform)$
setup_autoload("CONDENS/fenols.mac",fenols)$
setup_autoload("CONDENS/tracks.mac",tracks)$
setup_autoload("CONDENS/invert.mac",invert)$
setup_autoload("CONDENS/btriang.mac",btriang)$
setup_autoload("CONDENS/simu.mac",ff)$
setup_autoload("CONDENS/condens.mac",condens)$
print_true:false$
help(arg):=arg(help)$
```

CONDENS also contains some help facilities. The standard help functions in Macsyma are

$$\text{DISPLAY}(``\textit{fct-name}'')$$
$$\text{APROPOS}(``\textit{fct-name}'').$$

We added two *help* functions: **MENU** and **HELP**. **MENU** will display the list of all the functions the package contains. **HELP**(`''`*fct-name*`''`) returns an information text describing the function, its syntax, how to enter its arguments, and an example.

CONDENS Session

Here is how to start a session in CONDENS:

```
(c1) load("initfile.mac");
Batching the file initfile.mac
(d1)                            initfile.mac
(c2) condens();
***********************************************************************
*                                                                     *
*  Hello ! WELCOME to CONDENS :   CONtrol DEsign of Nonlinear Systems *
*                                                                     *
*  a MACSYMA package for the design of controllers for the output     *
*  tracking problem using differential geometric concepts.            *
*                                                                     *
*        by:              OUASSIMA AKHRIF                              *
*                                                                     *
*        under the supervision of  Prof. GILMER BLANKENSHIP           *
*                                                                     *
***********************************************************************

           f,g and h,whenever mentioned, always designate the nonlinear
           dynamics of the nonlinear system affine in control:
                     dx/dt = f(x) + g(x) u
                       y   = h(x)
           Type  MENU(), to get a list of all functions available in CONDENS.
(d2)                            done
(c3) menu();


HELP(fun-name): generates an information text describing the function.
JACOB(f):computes the Jacobian matrix of f.
LIE(f,g): computes the Lie brackets of the vector fields f and g.
ADJ(f,g,k): computes the k-th adjoint of f and g.
LIDEV(f,h): computes the Lie derivative of the real valued function h
            along the direction defined by the vector field f.
NLIDEV(f,h,k): computes the k-th Lie derivative of h along f.
BTRIANG(f,g): checks if the nonlinear system is in block triangular form
RELORD(f,g,h): computes the relative order of the scalar control system
            dx/dt=f(x)+g(x)u, y=h(x).
TRANSFORM(f,g): treats the feedback linearization problem,that is checks
                if the system is linearizable and solves for the nonlinear
                transformation.
FENOLS(): module that treats the output tracking problem using feedback
            linearization.
INVERT(f,g,h): finds the left inverse to the original nonlinear system.
TRACKS(): module that treats the output tracking problem using the left-
            inverse system.

(d3)                            done

(c4) help(relord);

 relord(f,g,h) : computes the relative order of the scalar nonlinear
                system:   dx/dt = f(x) + g(x) u
                            y   = h(x)
                f is entered in the form   f:[f1(x),....,fn(x)],
```

```
            example:(if n=2)        f:[x1,x2**2],
          g is entered in the form  g:[g1(x),....,gn(x)],
            example:(if n=2)        g:[x2,x1*x2],
          h is entered in the form  h:h(x),
            example:              h:x1+x2,

(d4)                              done
```

All functions not explicitly described in this section are already available in the basic Macsyma system. We recommend that the reader consult the *Macsyma Reference Manual* [11] for details.

3.2 Differential Geometric Tools in CONDENS

In this subsection, we describe the basic functions available in CONDENS. Since we are using concepts from differential geometry, we have to manipulate vector fields and differential forms. Manipulations with vector fields and forms are so systematic that they are very well suited for symbolic computations. The functions described in this section perform some straightforward differential geometric computations such as *Lie brackets* and *Lie derivatives*, and more complex computations such as *Kronecker Indices* or *relative degree* of nonlinear systems.

JACOB(f): computes the *Jacobian* of f, that is, returns the matrix:

$$\begin{pmatrix} \frac{\partial f_1}{\partial x_1} & \cdots & \frac{\partial f_1}{\partial x_n} \\ \vdots & \cdots & \vdots \\ \frac{\partial f_n}{\partial x_1} & \cdots & \frac{\partial f_n}{\partial x_n} \end{pmatrix}.$$

Example

For the following nonlinear system:

$$\begin{pmatrix} \dot{x}_1 \\ \dot{x}_2 \end{pmatrix} = \begin{pmatrix} x_1^2 \\ x_1 x_2 \end{pmatrix} + \begin{pmatrix} 2x_1 \\ 1 \end{pmatrix} \tag{5}$$

$$y = x_1^2 + x_2^2$$

```
(c1) f:[x1**2,x1*x2];

                                     2
(d1)                            [x1 , x1 x2]
(c2) jacob(f);
```

```
            df   [ 2 x1   0  ]
(d2)        -- = [            ]
            dx   [  x2    x1 ]
```

LIE(f,g): computes the *Lie brackets* of the vector fields f and g:

$$[f,g] = \frac{\partial g}{\partial x} f - \frac{\partial f}{\partial x} g.$$

ADJ(f,g,k): computes the kth adjoint of f and g:

$$ad_f^k g = [f, ad_f^{k-1} g]$$
$$ad_f^0 g = g.$$

Example

For system (5):

```
(c3) g:[2*x1,1];
(d3)                            [2 x1, 1]
(c4) lie(f,g);
                                      2
(d4)                  [f, g] = [- 2 x1 , - 2 x1 x2 - x1]
(c5) adj(f,g,2);
                            2
(d5)      ad (f, g) = [0, x1  (- 2 x2 - 1) - x1 (- 2 x1 x2 - x1)]
            2
```

LIDEV(f,h): computes the *Lie derivative* of the real valued function h along the direction defined by the vector field f: $\mathbf{L_f}h$.

NLIDEV(f,h,k): computes the kth *Lie derivative* of h along f:

$$L_f^k h = L_f \left(L_f^{k-1} h \right)$$
$$L_f^0 h = h$$

Example

For system (5):

```
(c6) lidev(f,h);
                                      2        3
(d6)                  lf(h) = 2 x1 x2  + 2 x1
(c7) nlidev(f,h,3);
```

```
                 2              2    2            2         3          3   2
(d7) lf (h) = x1   (2 x1 (2 x2  + 6 x1 ) + 8 x1 x2  + 12 x1 ) + 12 x1   x2
       3
(c8) ratsimp(%);
                                              3   2       5
(d8)                          lf (h) = 24 x1   x2  + 24 x1
                                3
```

Next, we have three more complicated functions. They use the
"sub-functions" LIE, ADJ, etc. to compute quantities such as
Kronecker indices or *relative degree* of nonlinear systems.

KRONECK(f,g): used for multivariable systems,where g represents the set of vector fields $g_1, \ldots, g_m$

This function is useful when the nonlinear system is trans-
formable to a linear controllable system in Brunovsky canonical form. It computes the *Kronecker indices* of the equivalent
linear system the original nonlinear system is transformed to.
It returns a set of numbers:

$$k_1 \geq k_2 \geq \cdots \geq k_m$$

$$k_1 + k_2 + \cdots + k_m = n.$$

BTRIANG(f,g): This function checks if the nonlinear system $\dot{x} =$
$f(x) + \sum_{i=1}^{m} g_i(x)u_i$ is in *block triangular form* (see [8] for a
definition). The argument g of the function represents the m
vector fields $g_1, \ldots, g_m$. The code is in Figure 3.

RELORD(f,g,h): computes the *relative order* (see [3] for definition) of the single-input single-output nonlinear system:

$$\dot{x} = f(x) + g(x)u$$
$$y = h(x).$$

Example

For system (5):

```
(c7) h:x1**2+x2**2;
                                          2    2
(d7)                                     x2  + x1
(c8) relord(f,g,h);
relative order of system = 1

(d8)                            done
```

```
/***************************************************************************/
/*      This function checks if the system is in block triangular form     */
/***************************************************************************/

btriang(arg1,[arg2]):=block([x,xl,u,ind],
                    if arg1=help then return((
print("Btriang(f,g) : checks if the multivarible system dx/dt=f(x)+g(x)u is "),
print("                in block triangular form.  "),
print("                f and g are entered in the form f:[f1(x),...,fn(x)],   "),
print("                                                g:[[g1(x)],..[gm(x)]], "),
print("                Example:(if n=2,m=2)          f:[x1,x2**2],          "),
print("                                                g:[[x1,1],[x2,x1]],  ")
,done)),
                    f:arg1,g:part(arg2,1),n:length(f),m:length(g),
                    x:makelist(concat('x,i),i,1,n),
                    ind:makelist(concat('ind,i),i,1,n),
                    ind[1]:0,d:1,p1:1,it:n,
                    u:makelist(concat('u,i),i,1,m),
                    xl:makelist([],i,1,n),xl[p1]:u,
                    for i:1 thru n do
                    equ[i]:f[i]+sum(g[j][i]*u[j],j,1,m),
                    for p:1 thru n/m do(l:0,los:0,
                        for i:n thru 1 step -1 do (
                    for id :1 thru d do
                    if i=ind[id] then (if i>1 then i:i-1
                    else (los:1,id:d)),
                    if los #1 then for r:1 thru length(xl[p]) do
                            if equ[i]#subst(0,xl[p][r],equ[i]) then (
                            l:l+1,cx[it]:x[i],it:it-1,
                            xl[p+1]:endcons(x[i],xl[p+1]),
                            ind[d]:i,if d=n then (i:1,p:n/m),d:d+1,
                            r:length(xl[p]))),
                    if l#m then
                    (print("system is not in block triangular form"),
                    resul:0,p:n/m)
                    else resul:1 ),if resul=1 then
                    print("system in block triangular form"),done)$
```

Figure 3. Code for BTRIANG(f,g).

3.3 Analysis Tools in CONDENS

Using the above functions, CONDENS contains two independent modules for analysis of nonlinear control systems: TRANSFORM and INVERT. They treat the two problems: feedback linearization and invertibility of nonlinear control systems, respectively.

TRANSFORM (f,g): TRANSFORM treats the feedback linearization problem presented in Section 2.1.

Given the nonlinear control system:

$$\dot{x} = f(x) + \sum_{i=1}^{m} g_i(x)u_i. \tag{6}$$

TRANSFORM proceeds in the following manner:

- Using LIE, ADJ, and KRONECK, it computes the *Kronecker* indices and checks the three necessary and sufficient conditions (see [4] and [5]) for feedback linearizability.

- If the system is feedback linearizable, TRANSFORM investigates properties of the nonlinear system which simplify computation of the transformation. For example, TRANSFORM checks to whether the system is in block triangular form (see [8] and [9]) (using the function BTRIANG), or if it is scalar and satisfies the more restrictive conditions presented in [2].

- In the more general case, TRANSFORM proceeds to solve for the transformation (change of coordinates and feedback), say T and its inverse T^{-1} by using the Macsyma function "`desolve`". This function tries to solve the set of n ordinary differential equations defining the transformation (see ([4] and [5]).[1]

---

[1] If the Macsyma function "`desolve`" fails, then TRANSFORM fails. It would be possible to augment this function to solve the PDE's numerically; however, we have not implemented these enhancements.

Example

Check if the following nonlinear system is feedback-linearizable to a controllable linear system and if so, find the F-transformation.

$$
\begin{pmatrix} \dot{x}_1 \\ \dot{x}_2 \\ \dot{x}_3 \\ \dot{x}_4 \\ \dot{x}_5 \end{pmatrix} = \begin{pmatrix} \sin(x_2) \\ \sin(x_3) \\ x_4^3 \\ x_5 + x_4^3 - x_1^{10} \\ 0 \end{pmatrix} + \begin{pmatrix} 0 & 0 \\ 0 & 0 \\ 1 & 0 \\ 0 & 0 \\ 0 & 1 \end{pmatrix} \begin{pmatrix} u_1 \\ u_2 \end{pmatrix}
$$

```
(c2) load("initfile.mac");
(d2)                           initfile.mac
(c3) f:[sin(x2),sin(x3),x4**3,x5+x4**3-x1**10,0];
                              3        3        10
(d3)           [sin(x2), sin(x3), x4 , x5 + x4  - x1 , 0]
(c4) g:[[0,0,1,0,0],[0,0,0,0,1]];
(d4)               [[0, 0, 1, 0, 0], [0, 0, 0, 0, 1]]
(c5) transform(f,g);


Hello,TRANSFORM tries to solve the problem:                                   #
             Given the non linear system:
                   dx
                   -- = f(x) + u  g (x) + u  g (x)
                   dt          2  2        1  1
find a non singular transformation that takes this system to a controllable
linear system:
                         dz
                         -- = b . v + a . z
                         dt
checking if the system is in block triangular form....
system not in block triangular form ==> trying the general method...
the system is multi-input ==> computing first the Kronecker indices of the equ#
ivalent controllable linear system
Kronecker indices are:
k[ 1 ]= 3
k[ 2 ]= 2
checking the first condition of transformability.....
checking the second condition of transformability...
checking the third condition of transformability....
span of c1 =span of c1 /C
span of c2 =span of c2 /C
All the conditions are satisfied
trying to construct (if possible) the transformation....
The new state variables are :
z[ 1 ]= x1
z[ 2 ]= sin(x2)
z[ 3 ]= cos(x2) sin(x3)
z[ 4 ]= - x4
                   3    10
z[ 5 ]= - x5 - x4  + x1

The new control variables are :
```

```
                            3                   2
v[ 1 ]= cos(x2) cos(x3) (x4  + u1) - sin(x2) sin (x3)
               2        3    10          9
v[ 2 ]= - 3 x4   (x5 + x4  - x1 ) + 10 x1  sin(x2) - u2

(d5)                              done
```

INVERT (f,g,h): Given the nonlinear control system:

$$\dot{x} = f(x) + \sum_{i=1}^{m} g_i(x)u_i \qquad (7)$$
$$y = h(x).$$

INVERT tries to check if this system is strongly invertible by investigating the relative order of the system. In case it is invertible (that is in case the relative order is finite), INVERT returns this relative order, furthermore, it computes the left-inverse to system (7) and returns: $A_{inv}(x), B_{inv}(x), C_{inv}(x), D_{inv}(x)$ where:

$$\dot{x} = A_{inv}(x) + B_{inv}(x)\hat{y}$$
$$u = C_{inv}(x) + D_{inv}(x)\hat{y}.$$

For an example of the use of INVERT, see the example on the design package TRACKS.

3.4   Control Design Packages in CONDENS

FENOLS and TRACKS are two design packages contained in CON-DENS. Their purpose is essentially to use, respectively, the two modules TRANSFORM and INVERT to design a control law that forces the output of a nonlinear system to follow some desired trajectory. They are both "user-friendly," asking progressively for all the data they need. They are capable of automatically generating, upon request, Fortran codes for simulation purposes.

FENOLS (): This package follows exactly the design scheme of Figure 1, that is:

- It takes as input the nonlinear dynamics $f, g_1, \ldots, g_m$, the desired path $x_d(t)$ and the desired eigenvalues for the linear regulator.

- It uses the package TRANSFORM to transform the system into a controllable linear one, then FENOLS uses the desired eigenvalues (entered as inputs) to design a linear feedback control (by pole placement). This control solves the output tracking problem for the equivalent linear system.

- If the user is interested in simulation results, all he has to do is answer "yes" to a question posed by FENOLS at the end: "Are you interested in simulation results?"

Tracking Control for an Industrial Robot

This is the example of a *basic industrial robot*. It has one rotational joint and a translational joint. Using FENOLS, we try to design the control or the values of the torques that make the joint angles follow some desired trajectories. The system can be described by a state space representation of the form:

$$
\begin{pmatrix} \dot{x}_1 \\ \dot{x}_2 \\ \dot{x}_3 \\ \dot{x}_4 \end{pmatrix} = \begin{pmatrix} x_2 \\ f_1(\mathbf{x}) \\ x_4 \\ f_2(\mathbf{x}) \end{pmatrix} + \begin{pmatrix} 0 & 0 \\ \frac{1}{m_R+m_L} & 0 \\ 0 & 0 \\ 0 & \frac{1}{k-m_R\ell x_1+(m_R+m_L)x_1^2} \end{pmatrix} \begin{pmatrix} u_1 \\ u_2 \end{pmatrix}
$$

where

$$
f_1(\mathbf{x}) = x_1 x_4^2 - \frac{m_R\ell}{2(m_R+m_L)} x_4^2
$$

$$
f_2(\mathbf{x}) = \frac{-2[(m_R+m_L)x_1 - m_R\frac{\ell}{2}]}{k - m_R\ell x_1 + (m_R+m_L)x_1^2} x_2 x_4
$$

$$
y_1 = x_1
$$

$$
y_2 = x_3
$$

with $m_R = m_L = \ell = 2, k = 5$.

We are interested in designing a tracking controller to track the trajectories:

$$
\begin{aligned}
y_1(t) &= exp(t) \\
y_2(t) &= t
\end{aligned}
$$

```
(c2) load("initfile.mac");
(d2)    initfile.mac
(c3) fenols();
```

```
Hello,FENOLS tries to solve the problem:                              #
        Given the non linear system:
                                m
                              ====
                    dx    \
                    -- = >      u  g (x) + f(x)
                    dt    /      i  i
                              ====
                              i = 1
find a non singular transformation that takes this system to a controllable li#
near system:
                            dz
                            -- = b . v + a . z
                            dt
enter dimension of the state space
4;
enter dimension of the control space
2;
enter the values of f in the form[f1(x),f2(x),....,fn(x)] followed by ,
[x2,x1*x4**2-x4**2/2,x4,(4*x2*x4-8*x1*x2*x4)/(4*x1**2-4*x1+5)];
enter g (x) in the form [ g      (x) .... g      (x) ]
       1                   1, 1             1, 4
[0,1,0,0];
enter g (x) in the form [ g      (x) .... g      (x) ]
       2                   1, 1             1, 4
[0,0,0,4/(4*x1**2-4*x1+5)];

Entering TRANSFORM :                                                  #
checking if the system is in block triangular form....
system in block triangular form
the transformation is easy to construct
The new state variables are:
z[ 1 ]= x1
z[ 2 ]= x3
z[ 3 ]= x2
z[ 4 ]= x4
The new control variables are:
                     2
            2      x4
v[ 1 ]= x1 x4  - --- + u1
                  2
        4 x2 x4 - 8 x1 x2 x4        4 u2
v[ 2 ]= -------------------- + ----------------
                2                     2
        4 x1  - 4 x1 + 5     4 x1  - 4 x1 + 5
enter the desired trajectory in the form [xd(1),...xd(n)]
[exp(t),exp(t),t,1];
enter the desired eigenvalues of the linear controller in the form [delta(1),.#
..,delta(n)] followed by a ,
[-2,-2,-2,-2];
The tracking controller is :


                      2                           t
            (2 x1 - 1) x4  - 2 (- 4 x3 - 4 x1 + 5 %e  + 4 t)
u(x,xd)[ 1 ]= - -------------------------------------------------
                                    2
```

```
                                        2                         t
u(x,xd)[ 2 ]= ((8 x1 x2 - 4 x2) x4 + 4 x1   (- 4 x4 - 4 x2 + 4 %e  + 5)

      t    t
        - 4 x1 (- 4 x4 - 4 x2 + 4 %e  + 5) + 5 (- 4 x4 - 4 x2 + 4 %e  + 5))/4

Are you interested in simulation results?(answer y or n)
y;
enter filename of fortran code(without adding'.f')
examp1;
enter initial time you would like the simulation to start from
0.0;
enter final time tf
5.0;
enter step size h
0.01;
enter initial condition in the form[xo[1],...xo[n]]
[0,0,0,0];

    dimension x( 4 ),dx( 4 ),datad(1000, 4 ),data(1000,
  1 4 ),u( 2 ),y( 4 )
c     set no of equations
    n= 4
    m= 2
          .           ||
          .           ||   Fortran code generated
          .           \/   and written to a file.

(d3)                      done
```

TRACKS (): This package follows the design scheme in Figure 2. It uses the module INVERT to generate the output tracking controller. It presents the same interaction facilities as FENOLS, that is, takes as input the nonlinear dynamics $f, g_1, \ldots, g_m, h_1, \ldots, h_m$ and the desired trajectory $y_d(t)$ and generates upon request a Fortran program for simulation purposes.

Example

For the example of the basic industrial robot, applying this second design method, we obtain:

```
(c2) load("initfile.mac");
(d2)    initfile.mac
(c3) tracks();
Hello,TRACKS tries to solve the problem:
          .    ||
          .    ||    "Entering Input data"
          .    \/
```

```
enter the desired trajectory you want the output of your system to track
yd[ 1 ](t)=
exp(t);
yd[ 2 ](t)=
t;

Entering INVERT...
relative order of system = 2
The inverse system to our non linear control system is :
                        dx
                        -- = binv(x) . yi(t) + ainv(x)
                        dt
                        u(t) = Cinv(x) + Dinv(x) . yi(t)


                          2    2
                        d y1  d y2
where         yi(t) = [----, ----]
                          2    2
                        dt    dt


              [ x2 ]
              [    ]
              [ 0  ]                          [            2      2 ]
              [    ]                          [    2 x1 x4  - x4    ]
   Ainv(x) = [    ]        Cinv(x) = [  - ---------------- ]
              [ x4 ]                          [          2          ]
              [    ]                          [                     ]
              [ 0  ]                          [ 2 x1 x2 x4 - x2 x4 ]

              [ 0  0 ]                        [ 1            0        ]
              [      ]                        [                       ]
   Binv(x) = [ 1  0 ]        Dinv(x) = [           2           ]
              [      ]                        [     4 x1  - 4 x1 + 5 ]
              [ 0  0 ]                        [ 0 ---------------- ]
              [      ]                        [           4           ]
              [ 0  1 ]
```

Checking if yd(t) is trackable by our system (right-invertibility)
yd(t) can be tracked by the output y(t)

```
  The control ud(t) that makes y(t) track yd(t) is:
                          2    2       t
                    2 x1 x4  - x4  - 2 %e
        ud(t)[ 1 ]= - ------------------------
                              2
        ud(t)[ 2 ]= 2 x1 x2 x4 - x2 x4
```

Are you interested in simulation results? (Answer y or n)
y;
enter filename of fortran code

```
               .   ||
               .   ||  generating the fortran code
               .   \/

(d3)                         done
```

4   Conclusions

Differential geometric tools have proved to be very useful for analysis
of nonlinear systems. Manipulations of vector fields and forms are essentially straightforward; however, doing the computations by hand,
in particular on higher-dimensional spaces, is hard since the amount
of labour grows rapidly and becomes rather tedious. Moreover, there
is every chance that errors will be introduced. By using a computer
algebra system like CONDENS, much valuable time can be saved
and design methods that rely on rather sophisticated mathematical
tools can be made available to a wider class of users.

If a reader is interested in using CONDENS, he or she should
contact one of the authors.

Acknowledgments

This research supported in part by NSF Grant CDR-85-00108.

We would like to thank C.I. Byrnes for suggesting this research
area and J.P. Quadrat for introducing us to the potential of Macsyma.

References

[1] Boothby, W. M., *An Introduction to Differentiable Manifolds
and Riemannian Geometry*, Academic Press, New York, 1975.

[2] Brockett, R. W., *Feedback Invariants for Nonlinear Systems*,
IFAC Congress, Helsinki, pp. 1115-1120, 1978.

[3] Hirschorn, R. M., "Invertibility of nonlinear control systems,"
SIAM J. Control Optim., Vol. 17, pp. 289–297, 1979.

[4] Hunt, R. L., Su, R., and Meyer, G., "Design for multi-input
systems," *Differential Geometric Control Theory*, edited by R.
Brockett, R. Millman and H. J. Sussman, Birkhauser, Boston,
pp. 268-298, 1983.

[5] Hunt, R. L., Su, R., and Meyer, G., "Global Transformations of Nonlinear Systems," *IEEE Trans. Aut. Control*, AC-28, No. 1, pp. 24–31, 1983.

[6] Jakubczyk, B., and Respondek, W., "On linearization of control systems," *Bull. Acad. Polon. Sci. Ser. Sci. Math. Astronom. Phys.*, Vol. 28, pp. 517–522, 1980.

[7] Krener, A. J., "On the equivalence of control systems and the linearization of nonlinear systems," *SIAM J. Control*, Vol. 11, pp. 670–676, 1973.

[8] Meyer, G., and Cicolani, L., "Application of nonlinear system inverses to automatic flight control design-system concepts and flight evaluations," *AGARDograph 251 on Theory and Applications of Optimal Control in Aerospace Systems*, P. Kent, ed., reprinted by NATO, 1980.

[9] Meyer, G., "The design of exact nonlinear model followers," *Proceedings of Joint Automatic Control Conference*, FA3A, 1981.

[10] Su, R., "On the linear equivalents of nonlinear systems," *Systems and Control Letters*, Vol. 2, No. 1, pp. 48–52, 1982.

[11] The Matlab Group Laboratory for Computer Science, *Macsyma Reference Manual*, M.I.T., Cambridge, Mass., 1983.

[12] Wonham, W. M., *Linear Multivariable Control: A Geometric Approach*, Springer-Verlag, New York, 1979.

[13] Akhrif, O., and Blankenship, G. L., "Using Computer Algebra for Design of Nonlinear Control Systems," Invited paper, ACC 87, 1987.

[14] Akhrif, O., and Blankenship, G. L., *Computer Algebra for Analysis and Design of Nonlinear Control Systems*, SRC TR-87-52, 1987.

[5] Hunt, K. J., and Meyer, G., "Stable inversion for nonlinear systems," *Automatica*, Vol. [illegible], [illegible], pp. [illegible], 1997.

[6] Isidori, A., *Nonlinear Control Systems*, [illegible], Springer-Verlag, Berlin, [illegible].

[7] Krener, A. J., "On the equivalence of control systems and the linearization of nonlinear systems," *SIAM J. Control*, Vol. 11, pp. [illegible], 1973.

[8] Meyer, G., and Cicolani, L., "Application of nonlinear system inverses to automatic flight control design — system concepts and [illegible]," [illegible].

Chapter 4

Vector Fields and Nilpotent Lie Algebras

Matthew Grayson and Robert Grossman

1 Introduction

This paper is concerned with the efficient symbolic computation of flows of ordinary differential equations of the form

$$\dot{x}(t) = F(x(t)), \qquad x(0) \in \mathbf{R}^N.$$

In other words, we want to write the corresponding trajectory $t \to x(t)$ in closed form. Of course, since this cannot be done for most differential equations, it is important to find a means of approximating a general flow by a flow which is explicitly integrable in closed form. In this paper we describe an infinite dimensional family of explicitly integrable flows E; in a companion paper [6] we give an algorithm which, given a general flow F, will, in some cases, find a flow E which is explicitly integrable in closed form and is close to the original flow F. More generally, one could imagine approximating a flow by a flow which is explicitly integrable in closed form, flowing for a time step, finding a new approximating flow, and continuing.

The type of integrable flows discussed in this paper arise in a very simple fashion. We describe the basic idea here and refer to [4] for the details. We assume that the differential equation

$$\dot{x}(t) = E(x(t)), \qquad x(0) \in \mathbf{R}^N,$$

has the form

$$
\begin{aligned}
\dot{x}_1(t) &= a_1 \\
\dot{x}_2(t) &= a_2(x_1) \\
\dot{x}_3(t) &= a_3(x_1, x_2) \\
&\ \ \vdots \\
\dot{x}_N(t) &= a_N(x_1, x_2, \ldots, x_{N-1})
\end{aligned}
$$

for polynomials $a_1, \ldots, a_N$. Since the first equation can be integrated in closed form to yield x_1 as a function of time, and the jth equation depends only upon the first through $(j-1)$th equations, the entire system can be explicitly integrated in closed form.

In many examples, the vector field E defining the flow has some additional structure so that it can be written as the sum of two or more pieces. For example, suppose E depends upon two controls $t \to u_1(t)$ and $t \to u_2(t)$ and can be written as $E = u_1(t)E_1 + u_2(t)E_2$. Alternatively, suppose that E can be written as the sum of two pieces $E_1 + \epsilon E_2$, where ϵ is a parameter. Observe that if E_1 and E_2 are vector fields with polynomial coefficients which are homogeneous of degree 1, then the flow generated by E is of the form above and, hence, is explicitly integrable in closed form. See Section 2 and [4] for more details.

Section 2 provides some background on vector fields, flows, and Lie algebras. Section 3 gives an algorithm for producing flows associated with free, nilpotent Lie algebras. Section 4 considers flows which are modeled by nilpotent Lie algebras satisfying relations "at a point." The appendix contains a brief description of the Maple code which we used to construct the examples.

Nilpotent approximations of this type have been used in control theory by Crouch [1] and [2], Hermes [8], [9], and [10], and Krener [13]; and in partial differential equations by Folland and Stein [3], Hörmander [11], Rockland [14], and Rothschild and Stein [15].

2   Vector Fields, Flows, and Lie Algebras

In this section, we define homogeneous derivations and the flows they generate. One of the interests of homogeneous derivations is that their flows are integrable explicitly in closed form. Homogeneous derivations may be classified by the algebraic properties of the (Lie) algebras they generate. In the next section, we give an algorithm which yields homogeneous derivations which generate a Lie algebra of special type, called a free, nilpotent Lie algebra. In Section 4, we describe the analogous algorithm for derivations which satisfy relations "at a point." For background and related material, see Goodman [5], Jacobson [12], and Varadarajan [16].

Let R denote the space of polynomial functions on $\mathbf{R}^N$ and let $E_1, \ldots, E_M$ denote vector fields on $\mathbf{R}^N$ with polynomial coefficients; that is, E_i is of the form

$$E_i = \sum_{j=1}^{N} a_i^j D_j,$$

where $a_i^j = a_i^j(x_1, \ldots, x_N) \in R$ are polynomials and $D_j = \partial/\partial x_j$. These are also called *derivations*, since they are derivations of the ring of polynomials R. Recall that a derivation E of a ring R is a map $E : R \longrightarrow R$ satisfying the identities

$$\begin{aligned}
E(a+b) &= E(a) + E(b) \\
E(ab) &= aE(b) + bE(a),
\end{aligned}$$

for elements a and $b \in R$. The *flow generated by the derivation* $E = \sum_{j=1}^{N} a^j D_j$, *with initial point* $x^0 = (x_1^0, \ldots, x_N^0) \in \mathbf{R}^N$ is defined to be the solution of the initial value problem

$$\begin{aligned}
\dot{x}_1(t) &= a_1(x_1, \ldots, x_N) \\
&\vdots \\
\dot{x}_N(t) &= a_N(x_1, \ldots, x_N),
\end{aligned}$$

$$x_1(0) = x_1^0, \quad \ldots, \quad x_N(0) = x_N^0.$$

Let $t \to x(t)$ denote the solution to this initial value problem. We say that the point $x^1 \in \mathbf{R}^N$ is reached in time s in case $x^1 = x(s)$. Notice that flows form a semigroup; that, is, if we can flow from $x^0 \in \mathbf{R}^N$ to the point x^1, and from x^1 to the point x^2, then we can flow from x^0 to the point x^2.

We are interested in *homogeneous derivations of weight m*. We define these informally; for a more formal definition, see [5]. First,

assign weights to the coordinate variables: say x_1 is of weight ρ_1,
..., x_N is of weight ρ_N. Second, assign weights to monomials, by
defining the weight of the monomial to be

$$\text{wt}(x_1^{e_1} \cdots x_N^{e_N}) = \rho_1{}^{e_1} + \cdots + \rho_N{}^{e_N}.$$

Third, a polynomial is called homogeneous of weight m in case it is
a sum of monomials, each of weight m. Fourth, define the weight of
the coordinate derivation to be

$$\text{wt}(D_j) = \partial/\partial x_j = -\rho_j,$$

and the weight of the derivation $a(x)D_j$ to be $\text{wt}(a(x)) - \text{wt}(D_j)$.
Fifth, a general derivation $E_i = \sum_{j=1}^{N} a_i^j D_j$ is of weight m in case it
is the sum of derivations, each of weight m. Sixth, the weight of the
second-order differential operator is defined as

$$\text{wt}(a(x)D_i D_j) = \text{wt}(a(x)) - \text{wt}(D_i) - \text{wt}(D_j),$$

and the weight of higher-order differential operators are defined in
an analogous fashion.

If E is a derivation of weight 1 on $\mathbf{R}^N$, then the flow

$$\dot{x}(t) = E(x(t)), \qquad (0) = x^0 \in \mathbf{R}^N$$

is explicitly integrable in closed form. This is easy to prove by induction, since the coefficient of $\partial/\partial x_i$, for example, depends only upon
variables which have already been computed.

Given M derivations $E_1, \ldots, E_M$, we define the *Lie algebra generated by* $E_1, \ldots, E_M$ to be the smallest real subspace of derivations
of R which is closed under the formation of Lie brackets

$$[E, F] = EF - FE;$$

this is denoted

$$\text{g}(E_1, \ldots, E_M).$$

The *length* of a Lie bracket is defined inductively as follows: the
generators all have length 1. If E is a Lie bracket of length less than
or equal to k and F is a Lie bracket of length less than or equal to
l, then $[E, F]$ has length less than or equal to $k + l$. The algorithms
in this paper use the length of a Lie bracket to define homogeneous

derivations in the following way. A vector space decomposition of $\mathbf{R}^N = V_1 \oplus \cdots \oplus V_r$ is defined by setting

$$V_l = \text{span } \{E : E \text{ is a Lie bracket of length } \leq l\}.$$

If $e_j, \ldots, e_{j'}$ is a basis of V_l, then each coordinate x_j in the dual basis of coordinates $x_j, \ldots, x_{j'}$ is defined to have weight l, and homogeneous derivations are defined as above.

A Lie algebra is called *nilpotent of step* r in case any Lie bracket of length greater than r is zero. Note that if $E_1, \ldots, E_M$ are homogeneous of weight 1, then the Lie algebra they generate is nilpotent of step r and any bracket of length l is homogeneous of weight l. This is because if F_1 is a derivation homogeneous of weight k and F_2 is a derivation homogeneous of weight l, then $[F_1, F_2]$ is homogeneous of weight $k + l$. A Lie algebra is *free* in case it satisfies as few relations as possible. Note that a Lie algebra always satisfies some relations, since, for example

$$[E_1, E_2] = -[E_2, E_1]$$

and

$$[[E_1, E_2], E_3] + [[E_2, E_3], E_1] + [[E_3, E_1], E_2] = 0.$$

Also, there will be additional relations that are a consequence of these. A Lie algebra is free in case any relation is a consequence of relations that are of the form above. Rather than try to make this precise, we give a basis for a free, nilpotent Lie algebra that is due to M. Hall [7].

Definition 2.1 *Each element of the Hall basis is a monomial in the generators and is defined recursively as follows. The generators $E_1, \ldots, E_M$ are elements of the basis and of length 1. If we have defined basis elements of lengths $1, \ldots, r - 1$, they are simply ordered so that $E < F$ if length$(E) < $ length(F). Also if length$(E) = s$ and length$(F) = t$ and $r = s + t$, then $[E, F]$ is a basis element of length r if:*

1. E and F are basis elements and $E > F$, and

2. if $E = [G, H]$, then $F \geq H$.

3   Free, Nilpotent Lie Algebras

In this section we provide motivation by reviewing an algorithm, whose input consists of a rank r, and whose output consists of two vector fields E_1 and E_2 on $\mathbf{R}^N$ which generate a Lie algebra isomorphic to the free, nilpotent Lie algebra $g_{2,r}$ on two generators of rank r. Here N is the dimension of the Lie algebra. The proof can be found in [4]. In the next section we consider the analogous problem when the Lie algebra satisfies relations. Recall that the flows of these vector fields can all be integrated explicitly in closed form.

Fix the rank $r \geq 1$ of the free, nilpotent Lie algebra $g_{2,r}$, and number the basis elements for the Lie algebra by the ordering from Definition 2.1, *i.e.*, $E_3 = [E_2, E_1]$, $E_4 = [E_3, E_1]$, $E_5 = [E_3, E_2]$, etc. Consider a basis element E_k as a bracket in the lower order basis elements, $[E_{i_1}, E_{j_1}]$, where $i_1 > j_1$. If we repeat this process with E_{i_1}, we get $E_k = [[E_{i_2}, E_{j_2}], E_{j_1}]$, where $j_2 \leq j_1$ by the Hall basis conditions. Continuing in this fashion, we get

$$E_k = [[\cdots[[E_{i_m}, E_{j_m}], E_{j_{m-1}}], \cdots, E_{j_2}], E_{j_1}],$$

where $i_m = 2$, $j_m = 1$, and $j_{n+1} \leq j_n$ for $1 \leq n \leq m - 1$. This defines a partial ordering of the basis elements. We say that E_k is a *direct descendant* of each E_{i_j}, and we indicate this by writing $k \succ i_j$.

Define monomials $P_{2,k}$ inductively by $P_{2,k} = -x_j P_{2,i}/(\deg_j P_{2,i} + 1)$, whenever $E_k = [E_i, E_j]$ is a Hall basis element, and where $\deg_j P$ is the highest power of x_j which divides P. If $l \succeq m$, then define $P_{l,m}$ to be the quotient $P_{2,m}/P_{2,l}$. Note that $P_{i,i} = 1$. The coefficients of these monomials guarantee that $-\frac{\partial}{\partial x_j} P_{i,l} = P_{k,l}$ whenever $i \prec k \prec l$. For example, if $E_k = [[[[[[E_2, E_1], E_1], E_1], E_2], E_4], E_4], E_7]$, and if $E_i = [[[E_2, E_1], E_1], E_1]$, then

$$P_{2,k} = -\frac{x_1^3 x_2 x_4^2 x_7}{3!2!}, \quad \text{and} \quad P_{i,k} = \frac{x_2 x_4^2 x_7}{2!}.$$

The following theorem summarizes the properties of the generators.

Theorem 3.1 *Fix $r \geq 1$ and let N denote the dimension of the free, nilpotent Lie algebra on 2 generators of rank r. Then the vector fields*

$$E_1 \;\; = \;\; \frac{\partial}{\partial x_1}$$

$$E_2 \;=\; \frac{\partial}{\partial x_2} + \sum_{i>2} P_{2,i}\frac{\partial}{\partial x_i}$$

have the following properties:

1. *they are homogeneous of weight one with respect to the grading*

$$\mathbf{R}^N = V_1 \oplus \cdots \oplus V_r,$$

   where $V_i =$ the span of Hall basis elements of length i.

2. *the Hall basis elements E_k they generate satisfy $E_k(0) = \frac{\partial}{\partial x_k}$;*

3. *the flow $E_1 + E_2$ is explicitly integrable in closed form;*

4. *the Lie algebra they generate is isomorphic to $g_{2,r}$.*

4   Relations at the Origin

In the previous section, we gave an algorithm which generates the explicitly integrable flows associated with free, nilpotent Lie algebras. Our eventual goal is to find as large a class as possible of explicitly integrable flows. In this section, we generalize the algorithm of the previous section by giving an algorithm which generates the explicitly integrable flows associated with nilpotent Lie algebras satisfying "relations at a point."

Let $\mathcal{E}_1,\ldots,\mathcal{E}_N$ denote the basis of the free, nilpotent Lie algebra on two generators $\mathcal{E}_1$ and $\mathcal{E}_2$ of rank r and dimension N, and let $E_1,\ldots,E_N$ denote the vector fields obtained by replacing each occurrence of the generators $\mathcal{E}_1$ and $\mathcal{E}_2$ in $\mathcal{E}_i$ by E_1 and E_2. Recall that a *homogeneous relation* is a relation of the form

$$\mathcal{E}_i = \sum_{j=1}^{N} \tilde{r}_{ij}\mathcal{E}_j,$$

where $\tilde{r}_{ij} = 0$, unless $\mathrm{wt}(E_i) = \mathrm{wt}(E_j)$, and $\tilde{r}_{ij}$ are scalars. We are interested in quotient algebras g defined by imposing homogeneous relations among the generators. Our eventual goal is to find a vector field model for an arbitrary quotient algebra g; that is, vector fields $E_1,\ldots,E_N$ corresponding to the Lie elements $\mathcal{E}_1,\ldots\mathcal{E}_N$ and satisfying the same bracket relations as the quotient algebra. The reason is simple: many questions about the structure of the algebra and about the associated dynamical system $\dot{x}(t) = E_1(x(t)) + E_2(x(t))$ are reduced to easily computed questions about the flows of the E_j's.

In this paper, we solve just a "local" version of this problem. To describe this problem, we fix the basis $\mathcal{E}_1, \ldots, \mathcal{E}_N$ in the domain, and the basis $D_1, \ldots, D_N$ in the range, and use the homogeneous relations to induce a map $R : \mathbf{R}^N \to \mathbf{R}^N$. The problem we solve is to find vector fields E_1 and E_2 such that the associated Lie brackets evaluated at the origin $E_1(0), \ldots, E_N(0)$ are the columns of the matrix R. We prove the theorem assuming the following technical hypothesis.

Assumption. The map R is the identity when restricted to the subspace spanned by the derivations $\mathcal{E}_i$ of weight less than or equal to $r/2$. In other words, the components r_{ij} of R satisfy $r_{ij} = \delta_{ij}$, whenever the weights of $\mathcal{E}_i$ and $\mathcal{E}_j$ are both less than $r/2$. Here δ_{ij} is the Kronecker delta.

Since the number of basis elements grows very quickly with the weight, this requirement affects very few of them. This restriction reduces the arguments essentially to the proof of the free case; under this condition, the images of the free nilpotent Lie algebra vector fields E_1 and E_2 under the map R generate the desired quotient algebra with relations at the origin. Indeed, the same map R takes the basis elements for the free nilpotent Lie algebra to a spanning set for the quotient algebra. Of course, in general, the operations of bracket and non-isomorphic linear mapping do not commute. With the above condition, it follows that they at least commute at the origin.

Let B denote the indices in the index set $\{1, \ldots, N\}$ which are of weight less than $r/2$; and let A denote the remaining indices. Also, let $\tilde{R} = (\tilde{r}_{ij})$ denote the matrix defining the homogeneous relations satisfied by the generators $\mathcal{E}_i$, and assume that

1. $\tilde{r}_{ij} = \delta_{ij}, \qquad$ for $i, j \in B$

2. for each fixed $i \in A$, $r_{ij} = 0$, for all $j \in A$ and $j \neq i$.

It is easy to see that, with these hypotheses, the matrix of relations $\tilde{R}$ is equal to the matrix R defined above using the bases $\mathcal{E}_i$ and D_j.

Define polynomials $P_{2,k}$ inductively by $P_{2,k} = -x_j P_{2,i}/(\deg_j P_{2,i} + 1)$, whenever $E_k = [E_i, E_j]$ is a Hall basis element, and where $\deg_j P$

is the highest power of x_j which divides P. If $l \succeq m$, then define $P_{l,m}$ to be the quotient $P_{2,m}/P_{2,l}$. Note that $P_{i,i} = 1$.

Theorem 4.1 *Choose $R = (r_{ij})$ satisfying the assumption above, and set*

$$E_1 = \frac{\partial}{\partial x_1}$$

and

$$E_2 = \sum_{i \geq 2} P_{2,i} \sum_{j=1}^{N} r_{ji} \frac{\partial}{\partial x_j}.$$

Then the Lie brackets E_j have the properties:

1. *they are homogeneous of weight 1 with respect to the grading*

$$\mathbf{R}^N = V_1 \oplus \cdots \oplus V_r,$$

   where V_i = the span of Hall basis elements of length i;

2. *the flows of the system*

$$\dot{x}(t) = E_1(x(t)) + E_2(x(t)), \qquad x(0) = x^0$$

   are explicitly integrable in closed form;

3. *at the origin, the vector fields satisfy*

$$E_i(0) = \sum_{j=1}^{N} r_{ji} \frac{\partial}{\partial x_j},$$

   for all $1 \leq i \leq N$, and so they satisfy the desired relations.

Proof of theorem. By construction the vector fields are homogeneous of weight 1; therefore their flows are explicitly integrable in closed form; hence, we need only prove assertion 3. As in the proof of the free case, this is proved using a lemma which is actually stronger than the theorem and which illuminates the structure of the vector fields E_j. We must introduce the minimum order of a polynomial which was defined in [4].

Definition 4.2 *Let A be any non-constant monomial in the variables $x_1, \ldots, x_n$. Define the minimum order of A by $m(A) = \min\{j : x_j | A\}$. If A is a polynomial with no constant term, then define $m(A)$ to be the maximum of the minimum orders of the monomials of A.*

Lemma 4.3 *If $E_k = [E_i, E_j]$, then*

$$E_k = \sum_{l \succeq k} P_{k,l} \sum_m r_{ml} \frac{\partial}{\partial x_m} + \sum_n Q_{k,n} \frac{\partial}{\partial x_n},$$

where $j \leq m(P_{k,l}) < k$, and $Q_{k,n}$ is a polynomial with no constant term with minimum order $< j$.

Assertion 3 follows easily from the lemma. The only coefficients which do not vanish at the origin are from the first term when $l = k$. This implies $E_k(0) = \sum_{m=1}^N r_{mk} \frac{\partial}{\partial x_m}$, as desired.

Proof of lemma. The proof is by induction. Direct computation yields

$$E_3 = \sum_{i \geq 3} P_{3,i} \sum_j r_{ji} \frac{\partial}{\partial x_j},$$

since $P_{2,i} = -x_1 P_{3,i}$ for all $i \geq 3$.

Consider the parents of E_k. By the inductive hypothesis:

$$E_i = \sum_{a \succeq i} P_{i,a} \sum_b r_{ba} \frac{\partial}{\partial x_b} + \sum_c Q_{i,c} \frac{\partial}{\partial x_c},$$

and

$$E_j = \frac{\partial}{\partial x_j} + \sum_{d \succ j} P_{j,d} \sum_e r_{de} \frac{\partial}{\partial x_e} + \sum_f Q_{j,f} \frac{\partial}{\partial x_f},$$

since the weight of $\mathcal{E}_j < \rho/2$, and so $E_j(0) = \frac{\partial}{\partial x_j}$.

Now $E_i = [E_p, E_q]$ and so $q \leq j$ by the basis condition. Therefore, we have the relations:

$$q \leq m(P_{i,a}) < i \quad \text{and} \quad m(Q_{i,c}) < q \leq j.$$

And trivially,

$$m(P_{j,d}) < j \quad \text{and} \quad m(Q_{j,f}) < j.$$

Now, the Lie bracket $[E_i, E_j]$ of E_i and E_j equals:

$$= \sum_{\substack{a \succeq i \\ d \succ j}} \sum_{\substack{b \\ e}} P_{i,a} r_{ba} r_{ed} \left(\frac{\partial}{\partial x_b} P_{j,d} \right) \frac{\partial}{\partial x_e} + \sum_{a \succeq i} \sum_{\substack{b \\ f}} P_{i,a} r_{ba} \left(\frac{\partial}{\partial x_b} Q_{j,f} \right) \frac{\partial}{\partial x_f}$$

$$+ \sum_{\substack{c \\ d \succ j}} \sum_{e} Q_{i,c} r_{ed} \left(\frac{\partial}{\partial x_c} P_{j,d} \right) \frac{\partial}{\partial x_e} + \sum_{c} \sum_{f} Q_{i,c} \left(\frac{\partial}{\partial x_c} Q_{j,f} \right) \frac{\partial}{\partial x_f}$$

$$- \sum_{a \succeq i} \sum_{b} r_{ba} \left(\frac{\partial}{\partial x_j} P_{i,a} \right) \frac{\partial}{\partial x_b}$$

$$- \sum_{c} \left(\frac{\partial}{\partial x_j} Q_{i,c} \right) \frac{\partial}{\partial x_c}$$

$$- \sum_{\substack{d \succ j \\ a \succeq i}} \sum_{\substack{e \\ b}} P_{j,d} r_{ed} r_{ba} \left(\frac{\partial}{\partial x_e} P_{i,a} \right) \frac{\partial}{\partial x_b} - \sum_{\substack{d \succ j}} \sum_{\substack{e \\ c}} P_{j,d} r_{ed} \left(\frac{\partial}{\partial x_e} Q_{i,c} \right) \frac{\partial}{\partial x_c}$$

$$- \sum_{\substack{f \\ a \succeq i}} \sum_{b} Q_{j,f} r_{ba} \left(\frac{\partial}{\partial x_f} P_{i,a} \right) \frac{\partial}{\partial x_b} - \sum_{f} \sum_{c} Q_{j,f} \left(\frac{\partial}{\partial x_f} Q_{i,c} \right) \frac{\partial}{\partial x_c}.$$

If A and B are any monomials with $m(A) < j$ then for any m, $A \frac{\partial}{\partial x_m} B$ either vanishes or has minimum order less than j. This, and the fact that $r_{ij} = 0$ unless E_i and E_j have the same weight, imply that the non-zero terms in the second, fifth, and sixth lines have minimum orders $< j$. If A is any monomial satisfying $m(A) < j$ and if $m \geq j$, then $\frac{\partial}{\partial x_m} A$ either vanishes or has minimum order less than j. This implies that the non-zero terms in the first and fourth lines have minimum orders $< j$. The remaining terms are

$$- \sum_{a \succeq i} \sum_{b} r_{ba} \left(\frac{\partial}{\partial x_j} P_{i,a} \right) \frac{\partial}{\partial x_b}.$$

When $a = i$, this term vanishes. If $m(P_{i,a}) = j$, then either $k = a$, in which case

$$- \sum_{b} r_{bk} \left(\frac{\partial}{\partial x_j} P_{i,k} \right) \frac{\partial}{\partial x_b} = \sum_{b} r_{bk} \frac{\partial}{\partial x_b},$$

or $a \succ k$, and

$$- \sum_{b} r_{ba} \left(\frac{\partial}{\partial x_j} P_{i,a} \right) \frac{\partial}{\partial x_b} = \sum_{b} r_{ba} P_{k,a} \frac{\partial}{\partial x_b}.$$

If $m(P_{i,a}) < j$, then $\frac{\partial}{\partial x_j} P_{i,a}$ is either zero, or it has minimum order less than j. If $m(P_{i,a}) > j$, then $\frac{\partial}{\partial x_j} P_{i,a} = 0$. We conclude that

$$- \sum_{a \succeq i} \sum_{b} r_{ba} \left(\frac{\partial}{\partial x_j} P_{i,a} \right) \frac{\partial}{\partial x_b} = \sum_{l \succeq k} \sum_{m} r_{ml} P_{k,l} \frac{\partial}{\partial x_l} + \sum_{g} Q_{k,g} \frac{\partial}{\partial x_g},$$

where $m(Q_{k,g}) < j$. This proves the lemma.

A   Appendix: Maple Code

The algorithms described in this paper were the result of computer experimentation using the symbolic packages Macsyma and Maple. In this appendix, we give examples of some of the very simple Maple code that we wrote. Figure 1 gives the Maple code to compute Lie brackets, and Figure 2 gives the Maple code to compute vector fields which satisfy relations at a point. This code is illustrated by computing two vector fields which satisfy the relation

$$E_6(0) + 2E_7(0) = E_8(0).$$

Finally, in Figure 3, we give the code which will flow along vector fields, and illustrate it by flowing along the sum of the two vector fields already computed.

Our example is a quotient algebra of $g_{2,4}$. This has a basis given by E_1, E_2 (the generators), $E_3 = [E_2, E_1]$, $E_4 = [E_3, E_1]$, $E_5 = [E_3, E_2]$, $E_6 = [E_4, E_1]$, $E_7 = [E_4, E_2]$, and $E_8 = [E_5, E_1]$. We impose the relation that

$$E_7(0) + 2E_8(0) = E_6(0).$$

The r matrix, followed by the two vector fields and their brackets is in Figure 4. The result of flowing along the $E_1 + E_2$ for time τ using the procedure in Figure 3 is displayed in Figure 5.

```
brac:= proc(v,w)
  #returns the Lie bracket of the vector fields v and w
  local temp,t,i,j,l;

  if not type(v,'vector') then
      ERROR('Not a vector')
  elif not type(w,'vector') then
      ERROR('Not a vector')
  fi;
  if vectdim(v)<>vectdim(w) then
    ERROR('Different dimensions!')
  fi;

  temp:=array(1..vectdim(v));
  t:=array(1..vectdim(v));

  for j from 1 to vectdim(v) do
    for i from 1 to vectdim(v) do
      t[i]:= v[i]*diff(w[j],x[i])-w[i]*diff(v[j],x[i])
    od;
    temp[j] := sum(t[l],l=1..vectdim(v));
  od;
  eval(temp);
end;
```

Figure 1. Maple code to compute Lie brackets.

```
HallRelations := proc(rho,maxdim)
  #rho is the maximum weight, maxdim is an upper bound for the dimension
  local left,right,i,j,a,b,w,w0,wptr,k,poly;

  left:= array(1..maxdim,sparse,[(3)=2]);
  right:= array(1..maxdim,[(3)=1]);
  poly:= array(1..maxdim,1..maxdim,sparse,[(3,1)=1]);
  w:=array(1..maxdim,sparse,[(1)=1,(2)=1,(3)=2]);
  wptr:=array(1..rho+1,sparse,[(1)=1,(2)=3,(3)=4]);
  #wptr(i) is the index of the first element of weight i.
```

Figure 2. Maple code to compute vector fields satisfying relations at point:
Part 1.

```
l:= 4:
print(e.3=[e.2,e.1]);
for w0 from 3 to rho do
                    #w0 is the weight of the produced brackets
   for i from round(w0/2) to (w0-1) do
                    #i is the weight of the left parent
     for j from wptr[i] to (wptr[i+1]-1) do
                    #j is the index of the left parent
       for k from wptr[w0-i] to (wptr[w0-i+1]-1) do
                    #k is the index of the right parent
         if k < right[j] then next fi;
                    #These are the Hall Conditions
         if j <= k then break fi;
         left[l]:= j;
                    #assign the left and right parents
         right[l]:= k;
                    #to the new bracket.
         print(e.l=[e.j,e.k]);
         w[l]  := w[j]+w[k];
                    #This is its weight.
         for m from 1 to k-1 do
                    #This loop assigns the multiindex for the coefficient.
           poly[l,m]:= poly[j,m] od;
         poly[l,k]:= poly[j,k]+1;
         l:= eval(l+1);
       od;
     od;
   od;
   wptr[w0+1]:= l;
od;
```

Figure 2. Maple code to compute vector fields satisfying relations at point: Part 2.

```
e1:=array(1..wptr[rho+1]-1,sparse,[(1)=1]):
e2:=array(1..wptr[rho+1]-1,sparse,[(2)=1]);
for i from 2 to rho do
                    #Sum over the weights
  for j from wptr[i] to wptr[i+1]-1 do
                    #Sum over elements of the same weight
    for q from wptr[i] to wptr[i+1]-1 do
      e2[j]:= eval(e2[j])+
          (product((-x[p])^poly[q,p]/(poly[q,p]!), p=1..right[q])*r[j,q])
    od;
  od;
od;
print(e1=eval(e1));
print(e2=eval(e2));
for i from 3 to wptr[rho+1]-1 do
  a:= left[i]:
  b:=right[i]:
  e.i:= brac(e.a,e.b);
  print(e.i=eval("));
od;
end;
```

Figure 2. Maple code to compute vector fields satisfying relations at point:
Part 3.

```
flow:=proc(v,p0)
  #flows returns the point reached after flowing for time tau
  #along the vector field v from the point p0

  local xt,r,x1,i,j,n;

  n := vectdim(v);
  xt := array(1..n);
  x1 := array(1..n);
  r  := array(1..n);
  r[1]:=v[1];
  for i from 1 to n-1 do
    xt[i] := int(r[i],t)+p0[i];
    r[i+1]:= subs( (x[j]=xt[j]) $ j=1..i,v[i+1]);
  od;
  xt[n] := int(r[n],t)+p0[n];
  for i from 1 to n do
    x1[i] := subs(t=tau,xt[i])
  od;
  eval(x1);
end;
```

Figure 3. Maple code to flow along a vector field.

```
r := array ( 1 .. 8, 1 .. 8,

                          [1, 0, 0, 0, 0, 0, 0, 0]

                          [0, 1, 0, 0, 0, 0, 0, 0]

                          [0, 0, 1, 0, 0, 0, 0, 0]

                          [0, 0, 0, 1, 0, 0, 0, 0]

                          [0, 0, 0, 0, 1, 0, 0, 0]

                          [0, 0, 0, 0, 0, 0, 0, 0]

                          [0, 0, 0, 0, 0, 1, 1, 0]

                          [0, 0, 0, 0, 0, 2, 0, 1]
)

>

HallRelations(4,8);
                          e3 = [e2, e1]

                          e4 = [e3, e1]

                          e5 = [e3, e2]

                          e6 = [e4, e1]

                          e7 = [e4, e2]

                          e8 = [e5, e2]
```

Figure 4. The relation matrix, the vector fields, and their brackets: part 1.

```
e1 = (array ( sparse, 1 .. 8, [1, 0, 0, 0, 0, 0, 0, 0]))

e2 = (array ( sparse, 1 .. 8,
```

$$[0,\ 1,\ -\ x[1],\ 1/2\ x[1]^2,\ x[1]\ x[2],\ 0,\ -\ 1/6\ x[1]^3\ -\ 1/2\ x[1]^2\ x[2],$$

$$-\ 1/3\ x[1]^3\ -\ 1/2\ x[1]^2\ x[2]\ ]))$$

```
e3 = (array ( 1 .. 8,
```

$$[0,\ 0,\ 1,\ -\ x[1],\ -\ x[2],\ 0,\ 1/2\ x[1]^2\ +\ x[1]\ x[2],\ x[1]^2\ +\ 1/2\ x[2]^2\ ]))$$

```
e4  = (array ( 1 .. 8,
```

$$[0,\ 0,\ 0,\ 1,\ 0,\ 0,\ -\ x[1]\ -\ x[2],\ -\ 2\ x[1]]]))$$

```
e5 = (array ( 1 .. 8, [0, 0, 0, 0, 1, 0, - x[1], - x[2]]))

e6 = (array ( 1 .. 8, [0, 0, 0, 0, 0, 0, 1, 2]))

e7 = (array ( 1 .. 8, [0, 0, 0, 0, 0, 0, 1, 0]))

e8 = (array ( 1 .. 8, [0, 0, 0, 0, 0, 0, 0, 1]))
```

Figure 4. The relation matrix, the vector fields, and their brackets: part 2.


```
>
  flow(add(e1,e2),array(1..8,sparse));

     array ( 1 .. 8,
     [tau, tau, - 1/2 tau , 1/6 tau , 1/3 tau , 0, - 1/6 tau , - 5/24 tau ])
>
  subs(tau=1,");

     array ( 1 .. 8, [1, 1, -1/2, 1/6, 1/3, 0, -1/6, -5/24])
```

Figure 5. The result of flowing along the vector field $e1 + e2$.

Acknowledgments

Matthew Grayson's research was supported in part by a National Science Foundation Postdoctoral Research Fellowship. Robert Grossman's research was supported in part by grant NAG2-513 from the NASA–Ames Research Center.

References

[1] Crouch, P. E., "Dynamical realizations of finite Volterra series," *SIAM J. Control Optim.*, Vol. 19, pp. 177–202, 1981.

[2] Crouch, P. E., "Graded vector spaces and applications to the approximations of nonlinear systems," to appear.

[3] Folland, G. B., and Stein, E. M., "Estimates for the $\bar{\partial}_b$ complex and analysis of the Heisenberg group," *Comm. Pure Appl. Math.*, Vol. 27, pp. 429–522, 1974.

[4] Grayson, M., and Grossman, R., "Models for free, nilpotent lie algebras," *J. Algebra,* to appear.

[5] Goodman, R. W., *Nilpotent Lie Groups: Structure and Applications to Analysis,* Lecture Notes in Mathematics, No. 562, Springer-Verlag, New York, 1976.

[6] Grossman, R., and Larson, R. G., "Solving nonlinear equations from higher order derivations in linear stages," *Adv. Math.*, to appear.

[7] Hall, M., "A basis for free Lie rings and higher commutators in free groups," *Proc. Amer. Math. Soc.*, Vol. 1, pp. 575–581, 1950.

[8] Hermes, H., "Control systems which generate decomposible Lie algebras," *J. Diff. Eqns.*, Vol. 44, pp. 166–187, 1982.

[9] Hermes, H., "Nilpotent approximations of control systems and distributions," *SIAM J. Control Optim.*, Vol. 24, pp. 731–736, 1986.

[10] Hermes, H., Lundell, A., and Sullivan, D., "Nilpotent bases for distributions and control systems," *J. Diff. Equations*, Vol. 55, pp. 385–400, 1984.

[11] Hörmander, L., "Hypoelliptic second order differential equations," *Acta. Math.*, Vol. 119, pp. 147–171, 1968.

[12] Jacobson, N., *Lie Algebras*, John Wiley and Sons, New York, 1962.

[13] Krener, A., "Bilinear and nonlinear realizations of input-output maps," *SIAM J. Control Optim.*, Vol. 13, pp. 827–834, 1975.

[14] Rockland, C., "Intrinsic nilpotent approximation," *Acta Applicandae Math.*, Vol. 8, pp. 213–270, 1987.

[15] Rothschild, L. P., and Stein, E. M., "Hypoelliptic differential operators and nilpotent groups," *Acta. Math.*, Vol. 37, pp. 248–315, 1977.

[16] Varadarajan, V. S., *Lie Groups, Lie Algebras, and their Representations*, Prentice-Hall, Inc., Englewood Cliffs, 1974.

Chapter 5

FIDIL: A Language for Scientific Programming

Paul N. Hilfinger and Philip Colella

1 Introduction

One fundamental goal of research in programming language design is to provide a better fit between problems and programming notation. In scientific computation, this quest is sometimes described as one of reducing the "semantic distance" between abstract mathematical descriptions of numerical methods and programs that implement them—in effect, of making abstract mathematical descriptions into programs. In its most ideal form, such a goal is generations beyond the current state of the art. However, there are intermediate points along the way to which we might aspire. In this paper, we describe one of them.

Currently, most numerical scientific programming is done in Fortran. This language has served its purpose well, but the basic operators, quantities, and definitional facilities that it supports are rather limited. Under the "Fortran model" of computation, programs consist of sequences of individual arithmetic operations on numbers contained in named scalar variables or in individual elements of arrays. Fortran provides a certain amount of abstraction in the form of subprograms, but these are sufficiently clumsy to use and define that in practice their application is limited (at least when measured against the practice in other programming languages). Despite these oft-

cited limitations, the scientific community has largely adapted itself
to Fortran, and has developed a large body of software in the form
of libraries and application code.

Modern advances in numerical algorithms—motivated both by
new applications and increased processing power—have led to increasingly complex programs, have made the task of converting algorithms to programs increasingly difficult, and thus have made it
increasingly attractive to automate this conversion. One approach
to automation is to provide a programming language that makes it
easier to write about more complex, "bigger" entities: about operators on arrays rather than on individual array elements, about index
sets with more general shapes than rectangles, and so forth. We
have taken just this approach with the FIDIL (FInite DIfference)
language, which we will describe in this article.

It is very difficult to evaluate a programming language, especially
in the very early stages of its deployment. In order to give the
reader some basis for judgment, we will present (in Section 4) several
realistic examples of FIDIL's use that are somewhat longer than is
traditional for an overview paper about a programming language.
We will start with a brief overview of FIDIL, first with a description
of features common to most programming languages (Section 2), and
then with a description of features specifically intended for scientific
computation (Section 3).

The FIDIL system is still under development. Our experimental
compiler is a Common Lisp program that runs on workstations and
translates FIDIL into Fortran programs for a Cray X-MP. The use
of Fortran as an intermediate language makes it relatively easy to
use existing Fortran libraries, and may ease the task of porting the
compiler to other machines.

2   Overview of General–Purpose Facilities

FIDIL, like any programming language built with a specific problem area in mind, comprises not only features specific to that area,
but also a more general framework such as one might find in any
language. We will begin by describing this framework.

2.1   General Program Structure

A FIDIL program consists of a sequence of declarations of constants
and variables. Variable declarations use the "*type list-of-variable-
names*" format of Fortran and the Algol family, as in

> **integer** x, y;

Constant declarations all have the form "**let** $s_1 = d_1, s_2 = d_2, \ldots$,"
which evaluates the d_i in order and defines the symbols s_i to have
those values. As illustrated in the following example, constants can
be ordinary scalar values, arrays (called *maps* in FIDIL; see Section 3), functions, or types.

> **let**
> $n = 3$,   /* *Scalar constant* */
> $V = $ [0, 0, 0, 1],   /* *Array constant* */
> $rad = $ **proc** (**real** *deg*) $->$ **real**: 0.017453293 * *deg*,
>   /* *A function* */
> *Point* $= $ **struct** [**real** x, y];
>   /* *A type* */

Declarations may also be nested inside function definitions, in which
case, as is usual for languages in the Algol family, they apply only
within that function and are computed once for each call of the function. To accommodate separate compilation, variables and certain
constants can be *imported*—defined externally—or *exported* for im-
port by other programs. In the case of variables, this capability
corresponds to COMMON blocks in Fortran. The effect of an exe-
cutable FIDIL program as a whole is defined as a call to a function
with the distinguished name *main*.

Constant declarations and assignment-free expressions play an
important role in FIDIL; the language has a distinct bias toward
their use in places where programmers in Fortran, Algol, C, or Pas-
cal might use sequences of assignments. It is not that such traditional
programming is any more difficult in FIDIL than in Fortran, for example, but rather that it is less necessary. The bias takes the form
of primitive operations for building arbitrarily complex objects in
single expressions. As we shall illustrate in Sections 3 and 4, the end
result of this bias is that the execution of a FIDIL program tends

to consist of a sequence of assignment statements in which the computation of the value to be assigned is very large; that is, there is a great deal of computation between assignment statements. There are two reasons this effect is desirable. First, at an abstract level, the scientific programmer deals with "large" operations on large objects: decompositions of matrices, applications of difference operators to all values on a grid. It is only an artifact of conventional programming languages—inherited from underlying machine architectures—that such operations ultimately must be written as individual operations on scalar variables. Second, from the concrete level of the compiler, the analysis and transformation necessary to take advantage of pipelined or parallel machine architectures is generally easier when assignment statements are minimized.

FIDIL requires that all named entities be declared and that each declared entity have a single type, which may be indicated explicitly, as for variable declarations, or inherited from the entity's definition, as for constants. These types include the usual scalar types—integers, reals, complex numbers, booleans, and characters,—domain and map types (which encompass arrays), functional types, and record types. Since scalar types in FIDIL do not differ significantly from their realization in other languages, we shall say no more about them, and concentrate instead on domain and map types in Section 3, functions and functional types in Section 2.3, and record types in Section 2.2.

2.2   Record Types

A record value (or variable), also called a record, is a collection of named values (or variables), called *fields*. Record types describe a class of records by giving the number, names, and types of the fields of all records in this class.

> **let**
> $\qquad$ *State* $=$ *struct* [**real** x, y, px, py, m];
>
> $\qquad$ *State S*;

The declarations above define *State* to be a record type whose values have four real fields, and then define S to be a *State* variable. S may be assigned to or passed as a parameter to subprograms, just as for objects of scalar types. The individual fields of S are accessible using *field selectors*, which are notated with a functional syntax as in the following example.

$$x(S) := px(S) \;/\; m * dt + x(S);$$

In keeping with previous comments about the infrequency of assignment, a FIDIL programmer might write the following instead.

$$S := State[\; px(S)/m * dt,\; py(S)/m * dt,$$
$$px(S) + fx*dt,\; py(S) + fy*dt,\; m(S)];$$

The right-hand side of this latter assignment statement illustrates the use of a *record constructor* to create an entire record at once.

2.3   Defining Functions and Operators

Earlier, we saw a definition of a simple function for converting degrees to radians.

```
let
    rad = proc (real deg) -> real: 0.017453293 * deg;
```

The syntax here is suggestive: it has the same form as the definition of a named constant, suggesting that the phrase to the right of the equals sign denotes a value in its own right. This is indeed the case; the expression defining the function *rad* is a *subprogram literal*. It has no name in isolation, but simply denotes "a function taking a single real parameter, call it *deg*, and returning the real value computed by the formula 0.017453293 * *deg*." The most common use for subprogram literals is in the context shown—as definitions of function names—but they are sometimes useful as anonymous function arguments to other subprograms and, as we shall see later, in defining functionals.

One prominent characteristic of mathematical notation is our tendency to reuse the same notation for multiple purposes. Programming languages present more opportunities for such reuse, since they tend to introduce mathematically artificial distinctions—as between "short real" numbers and "long real" numbers. FIDIL allows the *overloading* of notation so that a conventional or suggestive name may be used wherever it is appropriate. Hence, the definition of *rad* above may be extended to cover long real numbers as well.

> **extend**
>   *rad* = **proc (long real** *deg*) –> **long real:** 0.0174532935199433 ∗ *deg*;

The compiler determines the particular definition of *rad* to use by the context of its use.

Another characteristic of mathematical notation, as contrasted with many programming languages, is that function calls are notated not just with alphanumeric names, but also with other operators having a more varied syntax. To accommodate this, FIDIL allows the definition and overloading of infix (binary), prefix, and postfix operators as functions or procedures. We might, for example, extend addition to work on *State* variables, as defined above.

> **extend**
>   + = **proc** (*State* p1,p2) –> *State* :
>   **begin**
>   **let**
> $mc = m(p1)+m(p2)$;
>   **return** *State* [$(x(p1)*m(p1)+x(p2)*m(p2))/mc$,
> $(y(p1)*m(p1)+y(p2)*m(p2))/mc$,
> $px(p1)+px(p2)$, $py(p1)+py(p2)$,
> $m(p1)+m(p2)$];
>   **end**;

Besides showing the extension of '+' to *States*, this example illustrates a few minor points of syntax: the use of **begin** and **end** to provide a way of grouping several declarations and statements into a single statement or expression, and the use of the exit construct **return** to indicate the value of a function.

One common form of function definition defines one function as a specialization of another with certain parameter values fixed. For example, the following two declarations are identical. The second uses a *partial function closure* to abbreviate the definition.

```
    let
       f = proc (State p) -> Force: attraction(p0, p);
    let
       f = attraction(p0,?);
```

Here, we assume that the function *attraction* is previously defined to compute the contribution to the force (gravitational or whatever) on its second argument due to its first. The notation *attraction(p0,?)* denotes a function of one argument that uses *attraction* to compute its result, using $p0$ as the first argument.

2.4   Functionals

FIDIL has been designed to accommodate "functional" programming, in which the principal operations employed are the applications of pure (side-effect-free or global assignment-free) functions to structured data. As we shall see, this particular programming method makes heavy use of functions on functions.

Of course, most conventional programming languages, including Fortran, provide the ability to pass functions as arguments to other subprograms. FIDIL goes further and allows functions to be *returned* as well, and in general to be calculated with the aid of appropriate operators. As an example, consider the extension of the (by now much-abused) operator '+' to functions; the sum of two unary functions is a new unary function that produces the sum of these functions' values. It can be defined as follows.

```
    let
       UnaryFunction = proc (real x) -> real;
    extend
       + =
          proc (UnaryFunction f1, f2) -> UnaryFunction:      /* (1) */
             proc (real y) -> real: f1(y) + f2(y);           /* (2) */
```

The fragment above first defines *UnaryFunction* as a mnemonic synonym for

```
    proc (real x) -> real
```

which is, in isolation, a type describing values that are "procedures taking a single real arguments and returning a real result." Next, the subprogram literal giving the value of '+' indicates that '+' is a binary operator on unary functions $f1$ and $f2$—line (1)—and that its value is the subprogram that takes a real argument, x, and returns the sum of $f1$ and $f2$ at y—line (2).

2.5   Generic Subprograms

As it stands, the definition of '+' in Section 2.4 works only for functions on **real** values. A definition of precisely the same form makes perfect sense for functions of any numeric type, however. FIDIL provides a notation whereby a single *generic* subprogram declaration can serve essentially as the template for an entire family of specific subprogram declarations. Thus, we can generalize the addition of functions as follows.

```
extend
    + = proc ( proc (?T x) -> ?T  f1, f2) -> proc (?T x) -> ?T:
          proc (T y) -> T: f1(y) + f2(y);
```

Here, the notation '$?T$' indicates a pattern variable for which any type may be substituted. This definition of '+' applies to any pair of (unary) functions on the same type, T, producing another function on T. The resulting function uses whatever definition of '+' is appropriate for values of type T. The actual rules here are somewhat tricky, since it is possible in principle to have the definition of '+' on T differ from place to place in a program. For the purposes of this paper, we shall simply assume that this situation does not occur and not go into the specific rules governing the selection of '+', on the general assumption that an unhealthy preoccupation with pathologies makes for poor language design.

2.6   Standard Control Constructs

FIDIL's constructs for conditional and iterative execution differ only in syntax from those of other languages. Figure 1 illustrates both in two fragments showing a sequential and then a binary search. In each case, the search routine accepts a one-dimensional array with a least index of 0 and a value to search for in the array, returning

```
let
   search1 =
      proc (Vector A; integer x) -> integer:
         for i from [0 .. upb(A)] do
            if x = A[i] then return i;
            elif x > A[i] then return -1;
            fi;
         od,
   search2 =
      proc (Vector A; integer x) -> integer:
         begin
            integer i, j;
            i := 0;  j := upb(A);
            do
              if i >= j
                 then exit;
              else
                 let m = (i+j) div 2;
                 if A[m] >= x then j := m;
                 else i := m+1;
                 fi;
              fi;
            od;
            return
              if j < i then -1
              elif A[i] = x then i
              else -1
              fi;
         end;
```

Figure 1. Two searches: *search1* is linear and *search2* is binary.

either the index of the value in the array, or -1 if the value does not appear.

The **if-elif-else-fi** construct, taken directly from Algol 68, allows the programmer to indicate a sequence of conditions and the desired computations to be performed under each of those conditions. It may be used either as a statement, to indicate which of several imperative actions to take, or as an expression, to indicate which of several possible expressions to compute. As we shall see in Section 3.2, the conditional construct also extends to conditions that produce arrays of logical values, rather than single logical values.

The **do-od** construct indicates an infinite loop, which can be exited by an explicit **exit** or **return** statement (the latter causing exit from the enclosing subprogram as well.) A preceding **for** clause specifies an index set for the iterations. The fragment above illustrates a simple iteration by 1 through a range of integers. More general iterations are also possible. For example, one can iterate two variables over a rectangular set of integer pairs using the following construct.

> **for** (i, j) **from** $[1..N, 1..M]$ **do** ... **od**;

Here, the pairs are enumerated in row major order (j varies most rapidly). One can specify strides other than one, as in the following.

> **for** i **from** $[1..N]$ **by** 2 **do** ... **od**;

3 Domains and Maps

Two classes of data type, *domains* and *maps*, play a central role in FIDIL, because of their natural applications in algorithms that involve discretizing differential equations. Together, they constitute an extension of the array types universally found in other programming languages. An array in a language such as Fortran can be thought of as a mapping from some subset of $\mathbf{Z}^n$ (the set of n-tuples of integers) to some codomain of values. Unlike mappings that are defined as subroutines, arrays can be modified at individual points in the index set. The index set of a conventional array is finite, rectangular, and constant throughout the lifetime of the array. One typically denotes operations on arrays with dimension-by-dimension loops over the index set, indicating an action to be performed for each data value in the array as an explicit function of the index of that value.

Maps are the FIDIL data objects corresponding to arrays. Unlike conventional arrays, however, their index set need not be rectangular or fixed, and the primitive operations provided by FIDIL encourage the programmer to describe operations upon them with single expressions that deal with all their values at once, generally without explicit reference to indices. To accomplish this, the concept of array is split into that of a *domain*, which corresponds to an index set and contains tuples of integers, and of a *map*, which consists of a domain and a set of values, one for each element of the domain.

3.1   Domains and Maps with Fixed Domain

We use the notation **domain**[n] to denote the type of an n-dimensional index set. A variable declared

> **domain**[2] D;

can contain arbitrary sets of pairs of integers. A particular rectangular domain may be generated using a *domain constructor*, as in the following example, which sets D to the index set of an N by M Fortran array.

> $D := [\,1\,..\,N, 1\,..\,M\,]$;

Several standard set operations apply to domains; the operators '+', '-', and '*' denote union, set difference, and intersection, respectively. In addition, there are several operations, summarized in Table 1, that are appropriate for index sets.

For a domain D, the notation

> $[D]\ T\ \ X$;

declares a variable X that is a map with index set D and element values of type T. The type T (the *codomain*) may be any type; it is not restricted to scalar types. As a shorthand, a simple rectangular domain constructor may also be used to denote the domain, as in the following.

> $[1\,..\,10, 1\,..\,M\,]\ T\ Y$;

In both these cases, the domain is evaluated at the time the variable is declared and remains fixed throughout the variable's lifetime.

The precise domains of formal parameters to procedures need not be specified, but may be supplied by the actual parameters at the time of call, as in Fortran. For example, the following header is appropriate for a function that solves a system of linear equations $Ax = b$.

> **let**
>> *solve* = **proc** ([*2] **real** A;
>> [] **real** b; **ref** [] **real** x) : ...

The notation '[*2]' indicates that the index sets of A is a two-dimensional domain whose contents is given by the arguments at the time of call. The notation '[]' is short for '[*1].' The dimensionality of the domain may be supplied by the call as well. For example, the following header is appropriate for a function that finds the largest element in an array.

> **let**
> $$largest = \textbf{proc} \; (\; [*?n] \; \textbf{real} \; A \;) \; -> \; \textbf{real} : \; ...$$

Again, in all of these cases, the index set of the formal parameters so defined is taken to be fixed over the call.

3.2   Operations on Maps

FIDIL provides for ordinary access to individual elements of a map. If d is an element of the domain of a map A, then $A[d]$ is the element of A at index position d. However, FIDIL encourages the programmer to avoid the use of this construct and to try to deal with entire arrays at once. There is an extensive set of pre-defined standard functions and constructs for forming map-valued expressions. This in turn makes it easier for the compiler to make use of implementation techniques, such as vectorization or parallel processing, that process the entire array efficiently.

As for arrays in most modern languages, entire FIDIL maps may be assigned in a single operation, as in

> $A := map\text{-}valued \; expression;$

Likewise, constant map values may be defined in one declaration:

> **let** $C = map\text{-}valued \; expression;$

Finally, there are various ways to assign to only a portion of a map. The statement

> $A \; *:= \; E;$

for a map-valued expression E, assigns to only those elements $A[p]$ for which p is an element of both the domains of A and E.

The *map constructor* allows specification of the elements and index set of an entire map in one expression. The statement

$$V := [\, E_1, E_2, \ldots \,]$$

assigns to V a one-dimensional map such that $V[i] = E_i$. One may also use a rule to specify the elements of a map, as in the following.

$$V := [\, i \textbf{ from } domainOf(V) : f(i) \,];$$

This assigns to V a map whose element with index i is $f(i)$. The function $domainOf$ yields the domain of its argument (a map). In the case of a constant map, the "i **from**" phrase is superfluous and may be omitted, as in

$$[\, domainOf(V): \ 0.0 \,];$$

which yields a map that is identically 0 on the domain of V.

A programmer can specify maps as *restrictions* of other maps or as *unions* of a set of maps having disjoint domains. The following definitions first use the restriction operator, **on**, to cause *inner* to be a copy of the portion of A whose indices are between 1 and 99 and *rim* to contain the values of B for the other indices of A. Finally, they define C to have the same interior as A and border as B, using the map union operator, '$(+)$.'

> **let**
> $A = [\, p \textbf{ from } [0..100, 0..100] : g(p) \,],$
> $B = [\, p \textbf{ from } domainOf(A) : h(p) \,],$
> $inner = A \textbf{ on } [1..99, 1..99],$
> $rim = B \textbf{ on } (domainOf(A) - domainOf(inside)),$
> $C = A\ (+)\ B;$

One of the most important operations in FIDIL is the "apply all" functional, denoted by the postfix operator '$@$'. Suppose that g is defined as follows.

> **let**
> $g = \textbf{proc } (\textbf{real } x,y) -> \textbf{real} : \ldots;$

Then the expression '$g@$' denotes a function on maps whose codomain is **real** that yields a map whose codomain is **real**. Formally, for maps (or map-valued expressions) A and B, we have the following.

$$g@(A,B) \equiv [\, i \textbf{ from } domainOf(A) * domainOf(B) : g(A[i], B[i]) \,]$$

That is, $g@(A,B)$ yields a map whose domain is the intersection of the domains of the maps A and B and whose value at each point in that intersection is found by applying the pointwise operator g to the values of A and B at that point.

For notational convenience, the standard arithmetic operators, comparison operators, and certain others are already overloaded to operate on maps, so that, for example, $A+B$, denotes a map whose elements are the sums of corresponding elements of A and B. These operators are also overloaded to take a scalar as one operand and a map as the other, with the usual meanings. Thus, $2.0 * A$ is the result of multiplying each element of A by 2.0.

The conditional expression also extends automatically to take **logical**-valued maps as its conditional test. For example, the following statement defines $A[i]$ to be taken from B at all indices where $B[i]$ is non-negative, and otherwise from C.

> **let** $A = $ **if** $B >= 0$ **then** B **else** C **fi**;

The expression $B >= 0$ evaluates to a map from the domain of B to logical values. For those indices at which $B \geq 0$, A is defined to agree with B. For those indices at which C is defined and $B < 0$, A and C agree. Another way of expressing this is to use the standard function *toDomain*, which converts a **logical**-valued map to a domain containing exactly those indices at which the map has a true value:

> **let** $A = (B$ **on** $toDomain(B >= 0)) (+) (C$ **on** $toDomain(B < 0))$.

Many map operations of interest in finite-difference methods are functions of neighbors of a map element, and not just the element itself. To accommodate such operations, FIDIL extends the shift operations on domains (Table 1) to maps in the obvious way (see Tables 2 and 3). Thus, after the definition

> **let** $C = A<<[1,1]$;

the map C has the property that $C[[2,2]]$ is equal to $A[[1,1]]$, and in general, that

> $C[p] = A[p - [1,1]]$.

This last equivalence also illustrates the extension of arithmetic operators to maps: elements of a two-dimensional domain are pairs of integers, represented as one-dimensional maps of two integers such as p. The subtraction $p - $ [1,1] is therefore the result of subtracting 1 from each element of p.

The apply-all and shift operators allow the succinct description of difference operators. For example, consider a map defined to sample the values of a function f at $N + 1$ discrete points between 0 and 1.

> **let** *fhat* = [i **from** [0 .. N] : $f(i * (1.0/N))$];

Then a discrete approximation to the first derivative of f over this range is given by the following definition.

> **let** $df = (\textit{fhat} - (\textit{fhat} << [1]\,)) * N;$

Working this out by hand will reveal that

$$df\,[1] = (df\,[1] - df\,[0]\,)*N, \quad df\,[2] = (df\,[2] - df\,[1]\,)*N, \ldots$$

Because the domain of *fhat* does not contain $N + 1$ and that of *fhat* << [1] does not contain 0, the domain of *df* is [1..N].

3.3   Maps with Flexible Domains

Map variables need not have fixed domains. The following declaration indicates that the domain of *front*, a two-dimensional array of *States*, can vary over time.

> **flex** [*2] *State   front*;

One can set the domain of such a map directly, as in the following assignment.

> *domainOf* (*front*) := ...;

Alternatively, one can assign a map value with an arbitrary two-dimensional domain to *front* as a whole:

> *front* := [i **from** D : $h(i)$];

An important use of flexible array types is in specifying "ragged" arrays. For example, consider the type *BinGrid* defined as follows.

Table 1. Some standard operations on domains.

Expression	Meaning
valtype(D)	For a domain D: shorthand for the type of the elements contained in D. If D is a **domain**$[n]$ for $n > 1$, then **valtype**(D) is $[1..n]$ **integer**. For $n = 1$, **valtype**(D) is **integer**.
$D_1 + D_2$	Union of D_1 and D_2.
$D_1 * D_2$	Intersection of D_1 and D_2.
$D_1 - D_2$	Set difference of D_1 and D_2.
p **in** D	A logical expression that is true iff p (of type **valtype**(D)) is a member of D.
$lwb(D)$, $upb(D)$	For D a **domain**$[n]$: A value of type **valtype**(D) whose kth component is the minimum (lwb) or maximum (upb) value of of the kth components of the elements of D.
$shift(D, S)$, $D << S$	Where S is of type **valtype**(D) (or **integer** if $n = 1$) and n is the arity of D: the domain $\{d + S \mid d$ **in** $D\}$.
$shift(D)$	Same as $shift(D, -lwb(D))$.
$contract(D, S)$	The domain $\{d$ **div** $S \mid d$ **in** $D\}$.
$expand(D, S)$	The domain $\{d * S \mid d$ **in** $D\}$.
$accrete(D)$	The set of points that are within a distance 1 in all coordinates from some point of D.
$boundary(D)$	$accrete(D) - D$.

Expression	Meaning
NullMap(*n, type*)	The null *n*-dimensional map with codomain *type*.
domainOf(X)	The domain of map X.
toDomain(X)	{ p **from** *domainOf*(X) : X[p] }, where X is a **logical** map.
image(X)	Where X is a map whose codomain is of type [*n] **integer**: the domain $\{d \mid X[p] = d$ for some $p\}$.
upb(X), *lwb*(X)	*upb*(*domainOf*(X)), *lwb*(*domainOf*(X))
X * Y	For X and Y maps such that Y's codomain is **valtype**(*domainOf*(X)): a map object—assignable if X is assignable—such that $(X*Y)[p] \equiv X[Y[p]].$ This is the composition of X and Y; its domain is $\{p \in \mathrm{domainOf}(Y) \mid Y[p] \in \mathrm{domainOf}(X)\}$.
shift(X,S), $X<<S$ *shift*(X)	Where S is a [1..n] **integer** (an integer for $n = 1$), with default value -lwb(X), and n is the arity of X: the map X * [p **from** *domainOf*(X): $p-S$]. The operator *shift*, as well as *contract* and *expand* below, yields an object that is assignable if X is assignable.
contract(X,S)	X * [p **from** *contract*(*domainOf*(X), S): $S*p$].
expand(X,S)	X * [p **from** *expand*(*domainOf*(X), S): p/S].
X **on** D	The map X restricted to domain D. Also assignable if X is.
X (+) Y	Where *domainOf*(X) $\cap$ *domainOf*(Y) = {}: the union of the graphs of X and Y, whose codomains must be identical and whose domains must be of identical arity.

Table 2. Some standard operations on maps, part 1.

Expression	Meaning
$concat(E_1,\ldots,E_n)$	Concatenation of the E_i. There must be a type T such that each E_i is either a 1-dimensional map with a contiguous domain and codomain T, or a value of type T (which is treated treated as a one-element map with lower bound 0). The result has the same lower bound as E_1 and a length equal to the sum of the lengths of the E_i.
$F@$	Assuming that F takes arguments of type T_i and returns a result of type T: the (generic) function extending F to arguments of type $[D_i]\,T_i$ and return type D, where the D_i are domains of the same arity and D is the intersection of the D_i. The result of applying this function is the result of applying F pointwise to the elements corresponding to the intersection of the argument domains.
$F <@>$	For F as above returning type T_1: the extension of F to arguments of types $[D_i]T$ as above, returning a value of type $[D_1]T_1$ defined by $$\begin{aligned} &F{<}@{>}(x_1,\ldots,x_n) \\ &\quad = F@(x_1,\ldots,x_n)\ (+)\ (x_1\ \text{on}\ (D_1 - D)). \end{aligned}$$
$trace(A,S)$	where S is an integer i_1 or $S = [i_1,\ldots,i_r]$, A is a map with a rectangular domain of arity n, $0 < r < n$, and $1 \le i_1 < \ldots < i_r \le n$: The map B of arity $n - r$ defined as follows. $$B[j_1,\ldots,j_{i_1-1},j_{i_1+1},\ldots,j_n] = \sum_{k=lwb(A,i_1)}^{upb(A,i_1)} A[j_1,\ldots,j_{i_1-1},k,j_{i_1+1},\ldots,j_n].$$ That is, indices i_j are replaced by an index variable k and summed for each value of the other indices. To be well defined, the bounds of all the dimensions i_m of A must be identical. If $n - r = 0$, the result is a scalar.

Table 3. Some standard operators on maps, part 2.

```
let
      Unary = proc(?T x) -> ?T;

extend
    # =  proc(Unary f1,f2) -> Unary:         /* f1 composed with f2 */
              proc(T x) -> T: f1(f2(x)),
    + =  proc(Unary f1,f2) -> Unary:
              proc(T x) -> T: f1(x) + f2(x),
    * =  proc(Unary f1,f2) -> Unary:
              proc(T x) -> T: f1(x) * f2(x),
    * =  proc(Unary f; ?T a) -> Unary:        /* Scalar multiplication */
              proc(T x) -> T: f(x) * a,
    * =  proc(?T a; Unary f) -> Unary:
              proc(T x) -> T: a * f(x),
    Id = proc(?T f) -> ?T: f;                 /* Identity */

let
    /* Shift operators: for map A, (E1(k))(A) = A << [k,0] */
    E1 = proc(integer k) -> Unary: shift(? ,[k,0]),
    E2 = proc(integer k) -> Unary: shift(? ,[0,k]);

let
    Div = proc ([1..2] [*2] ?T x) -> [*2] ?T:
            (Id - E1(1))(x[1]) + (Id - E2(1))(x[2]);
```

Figure 2. A simple operator calculus.

```
let
      Bin = flex [] State;
      BinGrid = [D] Bin;
```

Each Bin has a one-dimensional domain that is independent of that
of any other Bin in a $BinGrid$. As its name might suggest, such a
data type might describe the state of a system of particles distributed
by spatial position into a set of bins.

3.4   An Illustrative Operator Calculus

Figure 2 displays a set of definition that we will use later. These
extensions of arithmetic and other operators allow the succinct description of new difference operators.

4   Examples

In this section, we will describe the implementation in FIDIL of
two sets of algorithms from numerical PDE's: particle methods for
the vorticity formulation of Euler's equations, and finite difference
methods for hyperbolic conservation laws. For each case, we will
present a simple algorithm, and then use that simple algorithm as a
building block for a more complex, but efficient algorithm for solving
the problem. We will restrict our attention to problems in two space
dimensions, since that is the case for which there are FORTRAN
implementations for all of these algorithms.

4.1   Example 1: Vortex Methods

Euler's equations for the dynamics of an inviscid incompressible fluid
in two space dimensions can be written in terms of transport of a
scalar vorticity ω as follows:

$$\frac{\partial \omega}{\partial t} + \mathbf{u} \cdot \nabla \omega = 0$$

$$\omega = \omega(\mathbf{x}, t) \in \mathbf{R}, \mathbf{x} \in \mathbf{R}^2$$

$$\mathbf{u} = K \star \omega = \int K(\mathbf{x} - \mathbf{x}')\omega(\mathbf{x}')d\mathbf{x}'$$

$$K = -\frac{\mathbf{x}^\perp}{2\pi|\mathbf{x}|^2} \ , \ \mathbf{x}^\perp = (-x_2, x_1).$$

The evolution of the vorticity ω is given by advection by a velocity
field $\mathbf{u}$ that is a non-local function of ω: $\mathbf{u} = K \star \omega = \nabla \times (\Delta^{-1}\omega)$.
The velocity $\mathbf{u}$ satisfies the incompressibility condition $\nabla \cdot \mathbf{u} = 0$, so
that the total vorticity in the system is conserved. If we consider
Lagrangian trajectories $\mathbf{x}(t)$ satisfying

$$\frac{d\mathbf{x}}{dt} = \mathbf{u}(\mathbf{x}(t), t), \tag{1}$$

then the vorticity along those trajectories remains unchanged:

$$\frac{d\omega}{dt}(\mathbf{x}(t), t) = \frac{\partial \omega}{\partial t} + \frac{d\mathbf{x}}{dt} \cdot \nabla \omega = \frac{\partial \omega}{\partial t} + \mathbf{u} \cdot \nabla \omega = 0,$$

i.e., $\omega(\mathbf{x}(t), t) = \omega(\mathbf{x}(0), 0)$.

Vortex methods use a particle representation of the vorticity as its fundamental discretization (see Chorin [4]).

$$\omega(\mathbf{x}, t) \approx \sum_j \omega_j f_\delta(|\mathbf{x} - \mathbf{x}_j(t)|), \quad f_\delta(\mathbf{x}) = \frac{1}{\delta^2} f\left(\frac{|\mathbf{x}|}{\delta}\right).$$

The function f_δ is a smoothed approximation to a delta function, with

$$2\pi \int_0^1 f(r) r\, dr = 1, \quad f(r) \equiv 0 \text{ for } r > 1.$$

The discretized dynamics are intended to mimic the Lagrangian dynamics of the vorticity given by (1). In semidiscrete form, they are given by the following.

$$\begin{aligned}
\frac{d\mathbf{x}_i}{dt} &= \mathbf{u}(\mathbf{x}_i) \\
\mathbf{u}(\mathbf{x}_i) &= \sum_j K_\delta(\mathbf{x}_i - \mathbf{x}_j)\omega_j \\
K_\delta(\mathbf{x}) &= K \star f_\delta(\mathbf{x}).
\end{aligned} \tag{2}$$

Since f_δ is a function of $|\mathbf{x}|$ only, explicit formulas can be given for K_δ, for example, when f is a polynomial. In any case, $K_\delta = K$ if $|\mathbf{x}| > \delta$.

Given the formula (2) for the velocity field evaluated at all the particle locations, straightforward application of some explicit ODE integration technique, which would call a procedure to evaluation $\mathbf{u}$, will yield a discretization in time. In Figure 3, we give a FIDIL program for evaluating the right-hand side of the ODE (2). The program is divided into two procedures. The first procedure *vortex_blob* evaluates the velocity field at an arbitrary point, $\mathbf{x} \in \mathbf{R}^2$: $\mathbf{u}(\mathbf{x}) = \sum_j K_\delta(\mathbf{x} - \mathbf{x}_j)\omega_j$. The cutoff function f_δ used here is due to Hald [5], and leads to a K_δ given by

$$\begin{aligned}
K_\delta(\mathbf{x}) &= \frac{\mathbf{x}^\perp}{2\pi\delta^2} F(\frac{|\mathbf{x}|}{\delta}), \text{ if } |\mathbf{x}| < \delta \\
&= -\frac{\mathbf{x}^\perp}{2\pi|\mathbf{x}|^2} \text{ otherwise}
\end{aligned}$$

where

$$F(r) = -36r^5 + 140r^4 - 196r^3 + 105r^2 - 14.$$

Vortex_blob takes as arguments x, the location where the velocity is to be evaluated, and *omega*, a one-dimensional map containing the information describing the vortices. The map takes values of type *vortex_record*, a user-defined record type containing *position*, *velocity*, and *strength* fields required to describe a single vortex. In writing *vortex_blob*, we have taken advantage of the fact that $F(1) = -1$ to compress the two cases in the definition of K_δ into a single expression. We use the operator *trace* to perform the sum in (2). Finally, the procedure *vorvel* uses *vortex_blob* to evaluate the velocity field induced at the locations of all the vortices. We use partial closure and function application to make the body of this procedure a single expression.

The principal difficulty with the algorithm described above is that the computational effort required to evaluate the velocities is $O(N^2)$, where N is the number of particles. Anderson [1] introduced a particle-particle, particle-mesh approximation to the $O(N^2)$ calculation, called the Method of Local Corrections (MLC). This algorithm is essentially linear in the number of particles, and does not introduce any additional loss of accuracy in the approximation to the underlying vorticity transport.

In the MLC algorithm, one introduces a finite difference mesh that covers the region containing the particles. Without loss of generality, we assume a square finite difference mesh covering the unit square, with mesh spacing $h = 1/M$ for some integer M, and satisfying $\delta < h$. We also assume that the particle locations $\mathbf{x}_j$ are all contained in a slightly smaller square $[Ch, 1 - Ch] \times [Ch, 1 - Ch]$, where $C \geq 1$ is in principle problem-dependent. In practice, satisfactory results have been obtained with $C = 2$. We also introduce the function $B : \mathbf{R}^2 \to \mathbf{Z}^2$ defined by $B(\mathbf{x}) = \mathbf{k}$ if

$$x \in [(k_1 - 1)h, k_1 h] \times [(k_2 - 1)h, k_2 h].$$

The algorithm is given as follows.

1) Compute $R : [1, \ldots, M]^2 \to \mathbf{R}^2$, a discrete approximation to $\Delta \mathbf{u}$ on the finite difference grid. If we denote by Δ^{fd} the 9-point discretization of Δ, then

```
external integer
   numvors;
external real
   twopi, delta;
let
   vortex_record =
       struct[[1 .. 2] real position, velocity; real strength],
   twoVector = [1 .. 2] real;

external [1 .. numvors] vortex_record vortices;

let vortex_blob = proc(twoVector x; [] vortex_record omega) -> twoVector:
begin

   let
       delx = x - position@(omega),

       distance_fcn = proc(twoVector x)->real:
          sqrt(x[1]**2 + x[2]**2),

       distance = distance_fcn@(delx),
       perp =  proc(twoVector x) -> twoVector: [-x[2] ,x[1]],
       maxrdel = max(?,delta)@(distance),
       minrdel = min(?,delta)@(distance),

       F = proc(real rd) -> real:
          (-14 + rd**2*(105 - rd*(196 - rd*(140 - 36*rd)))));

       return trace(
              F@(minrdel/delta) * perp@(delx) * strength@(omega)
              /(twopi*maxrdel**2),
          1);
end;

let vorvel = proc:
    velocity@(vortices) := vortex_blob(?,vortices)@(position@(vortices));
```

Figure 3. FIDIL program for velocity field evaluation.

$$(R^i)_\mathbf{k} = \omega_i(\Delta^{fd}K^i)_\mathbf{k}, \quad \text{if } |B(\mathbf{x}_i) - \mathbf{k}| \le C$$
$$= 0, \quad \text{otherwise}$$
$$R = \sum_i R^i$$

Here, $(K^i)_\mathbf{k} = K_\delta(\mathbf{k}h - \mathbf{x}_i)$, the array of values of $K_\delta(\cdot - \mathbf{x}_i)$ projected over the finite difference grid, and $|\mathbf{l} - \mathbf{m}| = \max(|l_1 - m_1|, |l_2 - m_2|)$, $\mathbf{l}, \mathbf{m} \in \mathbf{Z}^2$. We are able to truncate R^i to be zero outside a finite radius because $\Delta K_\delta(\cdot - \mathbf{x}_i) = 0$ outside the disc $|\mathbf{x} - \mathbf{x}_i| < \delta < h$; thus, sufficiently far from $\mathbf{x}_i$, the truncation error in Δ^{fd} applied to K_δ is small, and is well-approximated by zero.

2) Solve $\Delta^{fd}\mathbf{u}^{fd} = R$ using a fast Poisson solver. The use of fast Poisson solvers on rectangular grids is standard, and we won't discuss it here. The only subtlety in the present case is that the boundary conditions required are "infinite domain" boundary conditions, i.e., a set of discrete boundary conditions corresponding to the infinite domain Green's function $G \star \rho = u$, $G(\mathbf{x}) = -\frac{1}{2\pi}\log(|\mathbf{x}|)$. However, this can be done by performing two fast Poisson solves with Dirichlet boundary conditions.

3) Calculate the velocity field at the particle locations. For the collection of particles contained in a given cell, this is done in two steps. First, the velocity induced by particles in nearby cells is computed using the direct N-body formula (2). Then the effect of all more distant particle is interpolated from the finite difference grid, having first corrected the values from $\mathbf{u}^{fd}$ to reflect the fact that the influence of the nearby particles has already been accounted for in the local N-body calculation.

$$\mathbf{u}(\mathbf{x}_i) = \sum_{j:|B(\mathbf{x}_i)-B(\mathbf{x}_j)|\le C} \omega_j K_\delta(\mathbf{x}_i - \mathbf{x}_j) \tag{3}$$
$$+ I(\mathbf{x}_i; \tilde{\mathbf{u}}_{\mathbf{l}^1} \ldots \tilde{\mathbf{u}}_{\mathbf{l}^s})$$
$$\mathbf{l}^1, \ldots, \mathbf{l}^s = B(\mathbf{x}_i), \; B(\mathbf{x}_i) \pm [0,1], \; B(\mathbf{x}_i) \pm [1,0]$$
$$\tilde{\mathbf{u}}_{\mathbf{l}^s} = \mathbf{u}^{fd}_{\mathbf{l}^s} - \sum_{j:|B(\mathbf{x}_i)-B(\mathbf{x}_j)|} \omega_i K_\delta(\mathbf{l}^s h - \mathbf{x}_j)$$

Here, $I(\mathbf{x}; \tilde{\mathbf{u}}_{l^1} \ldots \tilde{\mathbf{u}}_{l^s})$ is calculated using complex polynomial interpolation in the plane.

$$
\begin{aligned}
I(\mathbf{x}) &= (\mathrm{Re}(\mathcal{I}(z)), -\mathrm{Im}(\mathcal{I}(z))) \\
\mathcal{I}(z) &= \sum_{q=0}^{4} a_q z^q, \quad z = (\mathbf{x}_1 + \sqrt{-1}\mathbf{x}_2)
\end{aligned}
$$

with the coefficients chosen such that

$$
\begin{aligned}
\mathcal{I}(z_s) &= \tilde{u}_{\mathbf{l}^s} - \sqrt{-1}\tilde{v}_{\mathbf{l}^s} \\
z_s &= (l_1^s + \sqrt{-1}l_2^s)h
\end{aligned}
$$

In Figure 4 we give a FIDIL implementation of Anderson's algorithm for evaluating the velocity field induced by a collection of vortices on themselves. The procedure MLC takes as input the arguments *psi*, which contains the vortex data structure, and h, the finite difference mesh spacing. *Psi* is a two-dimensional map whose values are one dimensional maps of varying sizes. The values of the one-dimensional maps are of type *vortex_record*. Thus *psi* represents the collection of vortices sorted into finite difference cells: the domain of *psi* is the finite difference grid, and a vortex at position $\mathbf{x}_i$ is represented by an entry in *psi*[k] only if $\mathbf{x}_i = \mathbf{k}$. MLC evaluates the velocity fields induced by the vortices in *psi* on themselves and stores them in *velocity* field of each *vortex_record* in *psi*.

The first step of MLC is performed in (A)–(B). For each cell, the velocity field induced by the vortices in that cell is calculated for all the points in a domain large enough so that the nine-point Laplacian applied to the resulting map is defined on D_C. To define the finite difference Laplacian, we use the operator and shift calculus defined in figure 2 which has been included in the header file *operator_calculus.h*. Then Δ^{fd} is applied, and the result is used to increment RHS. The second step the call to *PoissonSolve* (C), which we take to be an externally defined procedure, to obtain $\mathbf{u}^{fd}$ defined on the grid *domainOf(psi)*. The third step is performed in (D)–(E). For each cell $\mathbf{k}$, all of the vortices which will contribute to the sums in (3) for the vortices in *psi*[k] are gathered into a single map *psi_corrections*. Then the velocities of *psi*[k] are initialized with sum of the local N–body velocities. Finally, *u_fd_local*, the map containing the values of $\tilde{\mathbf{u}}$ are computed, and the interpolated velocities added to *velocity@(psi*[k]).

```
#include "operator_calculus.h"

let
    Fd_Values = [*2] twoVector;

let
    toComplex= proc(real a,b) -> complex: a + b*1i,
    iota = proc(domain[?n] D) -> [D] valtype(D): [i from D: i];

let PoissonSolve = proc(Fd_Values u) -> Fd_Values:
    external PoissonSolve;

let
    C = 2,
    D_C = [-C .. C, -C .. C],
    rhs_stencil = accrete(D_C),
    interpolation_stencil = Image([[0,0],[0,1],[1,0],[-1,0],[0,-1]]);

let MLC = proc(ref [*2] flex [] vortex_record psi):
begin
  let
    D = [i from domainOf(psi): length(psi[i]) /= 0];
  [domainOf(psi)] twoVector   RHS;

  RHS := [ domainOf(psi): [0.0, 0.0] ];                        /* (A) */

  let Laplacian = proc (Fd_Values u) -> Fd_Values:
        (4.*(E1(1) + E1(-1) + E2(1) + E2(-1)) - 20.0*Id +
        E1(1)#E2(1) + E1(1)#E2(-1) + E1(-1)#E2(1) + E1(-1)#E2(-1))(u)

  for k from D do
    let
        u_rhs_local = vortex_blob(?, psi[k])@(iota(rhs_stencil << k)*h);
      RHS := RHS + Laplacian(u_rhs_local)/(6.0*h**2)
  od;                                                          /* (B) */

  let u_fd = PoissonSolve(RHS,h);                              /* (C) */
```

Figure 4. FIDIL program for the Method of Local Corrections (Part 1 of 2).

```
for k from D do                                            /* (D) */
   flex [*] vortex_record psi_corrections;
   psi_corrections := NullMap(1,vortex_record);

   for j from D_C do
      psi_corrections := concat(psi_corrections, psi[k+j])
   od;

   velocity@(psi[k]) :=
      velocity@(psi[k]) + vortex_blob(?, psi_corrections)@(x@(psi[k]));
   u_fd_local :=
      u_fd << -k on interpolation_stencil
      - vortex_blob(?, psi_corrections)@
      (iota(interpolation_stencil << k)*h);
   velocity@(psi[k]) :=
      velocity@(psi[k])
      + interp(?, u_fd_local)@((position@(psi[k]) - k*h)/h);
   od;                                                     /* (E) */
end /* of MLC */;

let interp = proc(twoVector x; Fd_Values u_fd) -> twoVector:
begin
   let
      a = toComplex@(u_fd[][1],-u_fd[][2]),
      z = toComplex(x[1],x[2]),
      coef = [
         a[[0,0]],
         -(a[[1,0]] - a[[-1,0]] - 1i*(a[[0,-1]] - a[[0,1]]))*.25,
         (a[[-1,0]] + a[[1,0]] - a[[0,1]] - a[[0,-1]])*.25,
         -(a[[-1,0]] - a[[1,0]] + 1i*(a[[0,-1]] - a[[0,1]]))*.25,
         -(a[[0,0]] - a[[1,0]] - a[[-1,0]] - a[[0,1]] - a[[0,-1]])*.25
         ],
      pofz = coef[1] + z*(coef[2] + z*(coef[3] + z*(coef[4] + z*coef[5])));

   return([realPart(pofz), imagPart(pofz)]);
end /* of Interp */;
```

Figure 4. FIDIL program for the Method of Local Corrections (Part 2 of 2).

4.2   Example 2: Finite Difference Methods

A large class of time-dependent problems in mathematical physics can be described as solutions to equations of the form

$$\frac{\partial U}{\partial t} + \frac{\partial F}{\partial x} + \frac{\partial G}{\partial y} = 0$$

$$U = U(x,y,t) \in \mathbf{R}^M, F, G = F(U), G(U).$$

Such systems are known as conservation laws, since the equations are in the form of a divergence in space-time of (U, F, G). Hyperbolic refers to the fact that the underlying dynamics is given locally by the propagation of signals with finite propagation velocity; in particular, the initial value problem is well-posed. In general, F and G are nonlinear functions of U; for example, in the case of Euler's equations for the dynamics of an inviscid compressible fluid, if we denote the components of U by $U = (\rho, m, n, E)$, then the fluxes are given by

$$F(U) = (m, \frac{m^2}{\rho} + p, \frac{mn}{\rho}, \frac{m}{\rho}(E+p)),$$

$$G(U) = (n, \frac{mn}{\rho}, \frac{n^2}{\rho} + p, \frac{n}{\rho}(E+p)),$$

where the thermodynamic pressure p is given by

$$p = (E - \frac{m^2 + n^2}{2\rho^2})(\gamma - 1).$$

A widely used technique for discretizing conservation laws is to use finite difference methods whose form mimics at a discrete level the conservation form of the differential equations.

$$U_{i,j}^{n+1} = U_{i,j}^n + \frac{\Delta t}{\Delta x \Delta y}[\Delta y(F_{i+1/2,j} - F_{i-1/2,j})$$
$$+ \Delta x(G_{i,j-1/2} - G_{i,j+1/2})] \tag{4}$$

Here Δt is a temporal increment, Δx and Δy are spatial increments, and n, i, j, are the corresponding discrete temporal and spatial indices. The discrete evolution (4) has a geometric interpretation on the finite difference grid. We interpret $U_{i,j}^n$ as the average of U over the finite difference cell $\Delta_{i,j}$,

$$\Delta_{i,j} = [(i-1/2)\Delta x, (i+1/2)\Delta x] \times [(j-1/2)\Delta y, (j+1/2)\Delta y]$$

$$U_{i,j}^n \approx \frac{1}{\Delta x \Delta y} \int_{\Delta_{i,j}} U(x,y,n\Delta t)dxdy,$$

and the evolution of U can be thought of as given by a flux balance around the edges of $\Delta_{i,j}$.

The first algorithm we consider is a variation on one of the first algorithms for conservation laws, the Lax-Wendroff algorithm. We use a two-step formulation of a type first introduced by Richtmyer and Morton [6]. Here, and in what follows, we take the spatial grid to be square, i.e. $\Delta x = \Delta y = h$. The algorithm we will consider is given as follows.

$$
\begin{aligned}
U_{i+1/2,j+1/2} \;=\; & \frac{1}{4}(U_{i,j}^n + U_{i,j+1}^n + U_{i+1,j}^n + U_{i+1,j+1}^n) \qquad\qquad (5)\\[4pt]
+ \; & \frac{\Delta t}{4h}(F(U_{i,j}^n) + F(U_{i,j+1}^n) - F(U_{i+1,j}) - F(U_{i+1,j+1}^n))\\[4pt]
+ \; & \frac{\Delta t}{4h}(G(U_{i,j}^n) - G(U_{i,j+1}^n) + G(U_{i+1,j}) - G(U_{i+1,j+1}^n))
\end{aligned}
$$

$$F_{i+1/2,j} = \frac{1}{2}(F(U_{i+1/2,j-1/2}) + F(U_{i+1/2,j+1/2}))$$

$$G_{i,j+1/2} = \frac{1}{2}(G(U_{i-1/2,j+1/2}) + G(U_{i+1/2,j+1/2}))$$

It is clear from the above description that, if we want to evolve the solution on a finite rectangular grid for one time step, it suffices to provide additional solution values on a border of cells one cell wide all around the grid. A fairly general way to implement such boundary conditions is to provide a procedure $\phi(U,h)$ which returns U_B, the values required on the border of cells surrounding the grid where U is defined.

In Figure 5 we give a FIDIL implementation of the Lax-Wendroff algorithm (4), (5). Again, we use the operator and shift calculus from Figure 2 to implement the algorithm. We have split the implementation into two pieces. *LW_Flux* takes as input a map containing the values of U on the extended grid required to computed the fluxes

$F_{i+1/2,j}, G_{i,j+1/2}$. It returns a map of type [1 .. 2] **flex** *Values* containing those fluxes. This map-valued map is a natural type for describing fluxes for conservation laws, since one needs a map of type *Values* with a different domain for the flux in each coordinate direction. The procedure *LW* calls *phi* to calculate the boundary values *U_B*, calls *LW_Flux* with the first argument given by the direct sum of *U* and *U_B*. Finally, *U* is updated in place using (4).

In finite–difference calculations of solutions to hyperbolic conservation laws, it is often the case that the accuracy of the computed solution for a given rectangular mesh spacing can vary substantially as a function of space and time. If one wants to maintain a uniform level of accuracy in a calculation, it is necessary to vary the mesh spacing as a function of space and time, concentrating computational effort in regions where the error is largest. One approach is Adaptive Mesh Refinement (AMR) [3,2], in which the finite difference mesh is locally refined in response to some locally computed measure of the error. This leads to an algorithm in which the solution is defined on a hierarchy of rectangular grids, with the time evolution computed by multiple applications of a rectangular grid integration scheme such as the Lax-Wendroff algorithm described above. In addition, the error is periodically measured, and the grid hierarchy modified as required. In the following, we describe in detail the structure of the solution on the grid hierarchy, and give a FIDIL implementation of the time evolution of that solution for the special case of the Lax-Wendroff algorithm as the underlying integration scheme.

AMR is based on using a sequence of nested, logically rectangular meshes on which the PDE is discretized. For simplicity, we will also require that all the meshes be physically rectangular, with equal mesh spacing in both coordinate directions. We say a mesh at level l is a grid $D_{l,k}$ with mesh spacing h_l and define

$$D_l \; = \; \cup_k D_{l,k}.$$

The mesh spacings on the various grids are related by $h_l/h_{l+1} = r$ where the refinement ratio r is restricted to be an even number. By identifying a grid with the domain it covers, we have $D_1 = \cup_k D_{1,k} = D$, the problem domain. If there are several grids at level 1, the grid lines must align with each other, that is, each grid is a subset of a rectangular discretization of the whole space. We may often have overlapping grids at the same level, so that $D_{l,j} \cap D_{l,j'} \neq 0$, but how

```
#include "operator_calculus.h"
let
    nvar   = 4,
    Vector = [1 .. nvar] real,
    Values = [*2] Vector,
    Fluxes = [1 .. 2] flex Values,
    gamma = 1.4;
let LW_Flux = proc(Values U; real h,dt) -> Fluxes:
  begin
    let
        U_corner = 0.25*(E1(1) + Id + E2(1) + E1(1)#E2(1))(U),
        F_x = 0.5*((E2(1) + Id)#(E1(1) - Id))(F_fcn@(U)),
        G_y = 0.5*((E1(1) + Id)#(E2(1) - Id))(G_fcn@(U)),
        U_half = U_corner - dt*(F_x + G_y)/(2.*h);

    return [0.5*(Id + E2(1))(F_fcn@(U_half)),
            0.5*(Id + E1(1))(G_fcn@(U_half))]);
  end;
let LW = proc(ref Values U; real h,dt):
  begin
    let
        D = boundary(domainOf(U)),
        U_B = phi(U, D, h),
        Flux = LW_Flux(U (+) U_B, h, dt);
    U := U + dt*Div(Flux)/h;
  end;
let F_fcn = proc(Vector U) -> Vector:
  begin
  let p = (U[4] - (U[2]**2 + U[3]**2)/(2.*U[1]))*(gamma - 1.);
   return([U[2],U[2]**2/U[1] + p,U[2]*U[3]/U[1],U[2]*(U[4] + p)/U[1]]);
  end;
let G_fcn = proc(Vector U) -> Vector:
  begin
  let p = (U[4] - (U[2]**2 + U[3]**2)/(2.*U[1]))*(gamma - 1.);
   return([U[3],U[2]*U[3]/U[1],U[2]**2/U[1] + p,U[2]*(U[4] + p)/U[1]]);
  end;
```

Figure 5. FIDIL program for the Lax-Wendroff algorithm.

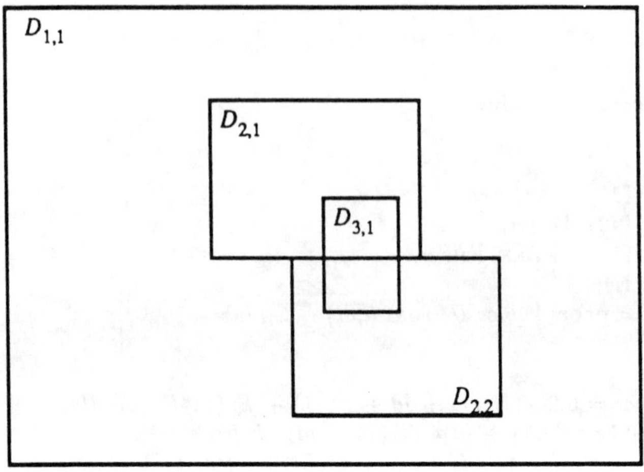

Figure 6. Grid hierarchy for three levels.

the grids intersect should have no effect on the solution. We require that the discrete solution be independent of how D_l is decomposed into rectangles. Grids at different levels in the grid hierarchy must be "properly nested." This means

(i) a fine grid is anchored at the corner of a cell in the next coarser grid.

(ii) There must be at least one level $l - 1$ cell in some level $l - 1$ grid separating a grid cell at level l from a cell at level $l - 2$, unless the cell abuts the physical boundary of the domain.

Note that this is not as strong a requirement as having a fine grid contained in only one coarser level grid.

Grids will be refined in time as well as space, by the same mesh refinement ratio. Thus,

$$\frac{\Delta t_l}{h_l} = \frac{\Delta t_{l-1}}{h_{l-1}} = \cdots = \frac{\Delta t_1}{h_1}$$

and so the same difference scheme is stable on all grids. This means more time steps are taken on the finer grids than on the coarser grids. This is needed for increased accuracy in time. In addition, the smaller time step of the fine grid is not imposed globally. Finally, $U^{l,k}$ denotes the solution on $D_{l,k}$; in addition, we can define U^l to be the solution on D_l, since the solution on overlapping grids at the same level are identical.

The AMR algorithm for advancing the solution on the composite grid hierarchy described above can be formulated as being recursive in the level of refinement. On a given level of refinement l, the algorithm can be broken up into three steps.

Step 1. Advance the solution on all the level l grids by one time step, using a conservative algorithm for doing so on a single rectangular grid. The only difficulty is in specifying the values along the outer border of $D_{l,k}$. For cells in that border contained in other grids at the same level, we copy the values from those other grids. For cells exterior to the physical domain, we use an appropriate variation of the physical boundary condition operator ϕ. For any remaining cells, we use values interpolated from the coarser levels. For the Lax-Wendroff algorithm described above, we can use a particularly simple interpolation scheme consisting of piecewise constant interpolation in space and linear interpolation in time using only the level $l-1$ grids. This is possible because Lax-Wendroff requires a border of boundary values that is only one cell thick, and because of the proper nesting requirement of the AMR grid hierarchy. After the solution is advanced, we use the numerical fluxes to initialize or update certain auxiliary variables used to maintain conservation form at the boundaries between coarse and fine grids; these quantities will be described in detail in Step 3.

Step 2. Advance the solution on all the level $l+1$ grids by r time steps, so that the latest values of the $l+1$ are known at the same time as the level l solutions obtained in step 1.

Step 3. Modify the solution values obtained in step 1 to be consistent with the level $l+1$ fine grid solutions. This will be the case if

(i) the grid point is underneath a finer level grid;

(ii) the grid point abuts a fine grid boundary but is not itself covered by any fine grid.

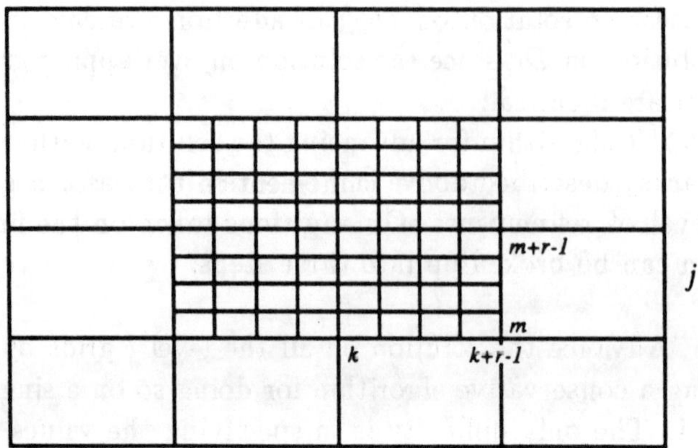

Figure 7. The coarse cell value is replaced by the average of all the fine grid points in that cell.

In case (i), the coarse grid value at level $l - 1$ is defined to be the conservative average of the fine grid values at level l that make up the coarse cell. After every coarse integration step, the coarse grid value is simply replaced by this conservative average, and the value originally calculated using (4) is thrown out. For a refinement ratio of r, we define

$$U_{i,j}^l := \frac{1}{r^2} \sum_{p=0}^{r-1} \sum_{q=0}^{r-1} U_{k+p,m+q}^{l+1}, \tag{6}$$

where the indices refer to the example in Figure 7.

In case (ii), the difference scheme (4) applied to the coarse cell must be modified. According to (4), the fine grid abutting the coarse cell has no effect. However, for the difference scheme to be conservative on this grid hierarchy, the fluxes into the fine grid across a coarse cell boundary must equal the flux out of the coarse cell. We use this to redefine the coarse grid flux in case (ii). For example, in the figure below, the difference scheme at cell (i,j) should be

$$\begin{aligned}
U_{i,j}^l(t + \Delta t_l) &= U_{i,j}^l(t) \\
&+ \frac{\Delta t_l}{h_l}[F_{i+1/2,j}^l(t)
\end{aligned}$$

$$-\frac{1}{r^2}\sum_{q=0}^{r-1}\sum_{p=0}^{r-1} F^{l+1}_{k+1/2,m+p}(t+q\Delta t_{l+1})]$$

$$+ \quad \frac{\Delta t_l}{h_l}[G^l_{i,j+1/2}(t)-G^l_{i,j-1/2}(t)]. \tag{7}$$

The double sum is due to the refinement in time: for a refinement ratio of r, there are r times as many steps taken on the fine grid as the coarse grid. If the cell to the north of (i,j) were also refined, the flux $G^l_{i,j+1/2}$ would be replaced by the sum of fine fluxes as well.

This modification is implemented as a correction pass applied after a grid has been integrated using scheme (4), and after the finer level grids have also been integrated, so that the fine fluxes in (7) are known. The modification consists of subtracting the provisional coarse flux used in (4) from the solution $U^l_{i,j}(t+\Delta t_l)$, and adding in the fine fluxes to according to (7). To implement this modification, we save a variable δF of fluxes at coarse grid edges corresponding to the outer boundary of each fine grid. After each set of coarse grid fluxes has been calculated in Step 1, we initialize any appropriate entries of δF with

$$\delta F_{i+1/2,j} = -F^l_{i+1/2,j}. \tag{8}$$

Since several coarse grids may overlap, it is possible that δF may be initialized more than once. However, since the coarse fluxes on overlapping cell edges for the same level are identical, the initial value so obtained is independent of which particular coarse grid flux is assigned last. At the end of each fine grid time step, we add to $\delta F_{i+1/2,j}$ the sum of the fine grid fluxes along the $(i+1/2,j)$th edge,

$$\delta F_{i+1/2,j} = \delta F_{i+1/2,j} + \frac{1}{r^2}\sum_{q=0}^{r-1} F^{l+1}_{k+1/2,m+q}. \tag{9}$$

Finally, after r fine grid time steps have been completed, we use $\delta F_{i+1/2,j}$ to correct the coarse grid solution so that the effective flux is that of (7). For example, for cell $(i+1,j)$, we make the correction

$$U^l_{i+1,j} := U^l_{i+1,j} + \frac{\Delta t_l}{h_l}\delta F_{i+1/2,j}. \tag{10}$$

If the cell $i+2,j$ were refined, we would also make the correction

$$U^l_{i+1,j} := U^l_{i+1,j} - \frac{\Delta t_l}{h_l}\delta F_{i+3/2,j},$$

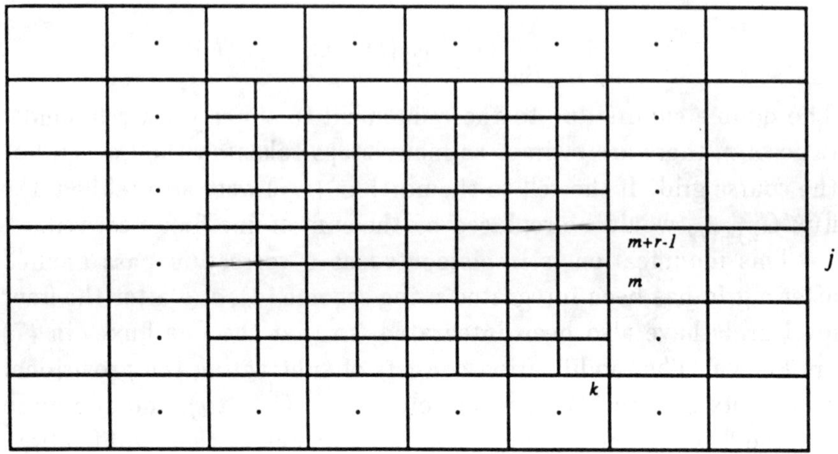

Figure 8. The difference scheme is modified at a coarse cell abutting a fine grid.

and similarly for the vertical fluxes. At the end of a time step, we
may have several fine grids available to update a given coarse cell
edge, since overlapping grids are permitted. For this reason, one
must keep track of the edges of a coarse cell that have already been
updated, and only perform the update once for each edge. As before,
it doesn't matter which fine grid actually performs the update for
any given edge, so the result is independent of the order in which
the fine grids are traversed.

In Figure 9, we give a FIDIL implementation of the integration
step outlined above for AMR. The main procedure *Step* takes a single integer argument l, the grid level being integrated. The principal variables are two copies U, U_new, of the composite map structure containing the entire set of solution values, with $U^{l,k}$ stored in
$U[l][k]$. The two sets of values U, U_new correspond to two different time levels, with times $time[l]$, $time[l] + delta\_t[l]$ depending
on the grid level l. We also define *Boundary_Flux*, in which the correction fluxes δF, δG, are stored for the outer edges of all the grids.
The refinement ratio s denoted by *nrefine*.

The first step of the AMR integration procedure is performed in
(A)–(B). For each level l grid solution $U^{l,k}$, appropriate boundary

conditions are calculated and stored in U_B, a map whose domain is $Boundary(domainOf(U[l][k]))$. In the first loop of this section, boundary values are interpolated from all possible $l-1$ grids, using piecewise constant interpolation in space, and linear interpolation in time. Since the domain of U_B is only one cell wide, proper nesting guarantees that all the values of U_B corresponding to points in the problem domain D will be set by this procedure. In the second loop, all of the values of U_B for which level l values are available are overwritten with them. Then the physical boundary condition procedure ϕ is called to fill in any remaining cells which extend outside D. Having obtained appropriate boundary values for $U^{l,k}$, we compute fluxes using *LW_Flux* and compute *U_new* using (4). Finally, we set *Boundary_Flux*$[l]$ and *Boundary_Flux*$[l+1]$ using (8) and (9) along the outer edges of $D_{l,k}$. The procedure *Project_Flux*, defined at the end of the figure, calculates the average of the fluxes in the right-hand side of (9).

The second step is performed at (C), with *Step* called recursively *nrefine* times with argument $l+1$.

The third step is in (D)–(E). In the first part of the loop over *Grids*, level l cells are incremented using the refluxing algorithm (10). The domains *D_fix* are used to keep track of which edges are being updated, so that no edge gets updated more than once. The final loop over *Gridsp* overwrites the level l values with the averages of the level $l+1$ values using (6).

```
export Step;

let
    maxlev  = 3,
    nrefine = 4,
    refine  = [nrefine,nrefine],
    Box     = [0 .. nrefine − 1, 0 .. nrefine − 1];

let
    Levels  = [1 .. maxlev];

let e_vector =
        proc (integer i, ndim) −> [1 .. ndim] integer:
            [j from [1 .. ndim] : if i = j then 1 else 0 fi ];

postfix operator ^*;
let
    ^* =
        proc(domain[?ndim] D) −> [1 .. ?ndim] domain[?ndim]:
            [i from [1 .. ndim] : D + D << e_vector(i, ndim) ],
    del =
        proc(domain[?ndim] D) −> [1 .. ndim] domain[ndim]:
            [i from [1 .. ndim] :
                (D + D << e_vector(i,ndim)) − (D − D << e_vector(i,ndim))];


external [Levels] flex [] flex Values  U, U_new;
external [Levels] flex []  Fluxes      Boundary_Flux;
external [Levels] real                 delta_t, h, time;
```

Figure 9. FIDIL program for Adaptive Mesh Refinement (Part 1 of 4).

```
let Step = proc (integer l) :
begin                                                          /* (A) */
    U[l] := U_new[l];
    alpha = (time[l-1] - time[l])/delta_t[l-1]

    let
        Grids = domainOf(U[l]),
        Gridsp = if l < maxlev then domainOf(U[l+1]) else NoGrids fi,
        Gridsm = if l > 1 then domainOf(U[l-1]) else NoGrids fi;

    for k from Grids do
        [Boundary(domainOf(U[l][k]))] Vector U_B;

        for km from Gridsm do
            for i from Box do
                contract(U_B << i, refine) *:=
                    alpha*U[l-1][km] + (1. - alpha)*U_new[l-1][km]
            od;
        od;

        for kb from Grids do U_B *:= U[l][kb] od;

        U_B *:= phi(U,domainOf(U_B),h[l]);

        let
            Flux = LW_Flux(U[l][k] (+) U_B, delta_t[l], h[l]);

        U_new[l][k] := U[l][k] + delta_t[l]*Div(Flux)/h[l];
        if l > 1 then
            Boundary_Flux[l][k] *:= Boundary_Flux[l][k] + Project_Flux(Flux);
        fi;
        for kp from Gridsp do
            Boundary_Flux[l+1][kp] *:= Flux*nrefine**2
                on del(contract(U[l+1][kp], refine))
        od;
    od;
```

Figure 9. FIDIL program for Adaptive Mesh Refinement (Part 2 of 4).

```
    time [l] := time [l] + delta_t [l];                    /* (B) */

if l < maxlev then

    for n from [1 .. nrefine] do Step(l+1) od;             /* (C) */

    for kp from Gridsp do                                  /* (D) */
        Boundary_Flux [l+1] [kp] :=
            delta_t [l] * Boundary_Flux [l+1] [kp] / (h * nrefine ** 2)
    od;

    for k from Grids do

        [1 .. 2] domain[2] D_fix;

        D_fix := domainOf( U [l] [k] ) ^*;

        for dir from [1 .. 2] do

            let E = e_vector(dir, 2);

            for kp from Gridsp do
                U_new [l] [k] *:=
                    U_new [l] [k]
                    + (Boundary_Flux [l+1] [kp] [dir] on D_fix [dir]);
                U_new [l] [k] << -E *:=
                    U_new [l] [k] << -E
                    - (Boundary_Flux [l+1] [kp] [dir] on D_fix [dir]);
                D_fix [dir] :=
                    D_fix [dir] - domainOf(Boundary_Flux [l+1] [kp] [dir])
            od;

        od;
```

Figure 9. FIDIL program for Adaptive Mesh Refinement (Part 3 of 4).

```
        for kp from Gridsp do

            U_new [l] [k] *:= [contract( U_new [l+1] [kp], refine): 0.0];

                for i from Box do
                    U_new [l] [k] *:=
                        U_new [l] [k]
                          + contract( U_new [l+1] [kp] << i, refine)
                od;


            od;
        od;
    fi;
end /* of Step */;                                         /* (E) */


let Project_Flux = proc(Fluxes Flux) -> Fluxes:
begin

    Fluxes OutFlux;

    domainOf@(OutFlux) := contract@(Flux, refine);

    for dir from [1 .. 2] do

        OutFlux[dir] := [domainOf(OutFlux[dir]): [ [1 .. nvar]: 0.0 ] ];
        let E = e_vector(if dir = 1 then 2 else 1 fi, 2);
        for i from [0 .. nrefine - 1] do
            OutFlux[dir] := OutFlux[dir] + contract(Flux[dir] << (i*E),refine) ;
        od;

    od;
    return(OutFlux);
end /* of Project_Flux */;
```

Figure 9. FIDIL program for Adaptive Mesh Refinement (Part 4 of 4).

Acknowledgments

Paul Hillfinger's research was supported in part by the NSF under grant DCR–8451213 and by Cray Research. Phillip Colella's research was supported by the U.S. Department of Energy, Office of Energy Research Applied Mathematical Sciences Program at the Lawrence Livermore National Laboratory under contract W–7405–Eng-48.

We would like to thank Luigi Semenzato for his comments on various versions of this document.

References

[1] Anderson, C. R., "A method of local corrections for computing the velocity field due to a distribution of vortex blobs," *J. Comput. Phys.*, Vol. 62, pp. 111–123, 1986.

[2] Berger, Marsha J. and Colella, Phillip, "Local adaptive mesh refinement for shock hydrodynamics," Lawrence Livermore National Laboratory, UCRL-97196 (1987), to appear in *J. Comput. Phys.*

[3] Berger, Marsha J. and Oliger, J., "Adaptive mesh refinement for hyperbolic partial differential equations," *J. Comput. Phys.*, Vol. 52, pp. 484–512, 1984.

[4] Chorin, Alexandre J., "Numerical Study of Slightly Viscous Flow," *J. Fluid Mech.*, Vol. 57, pp. 785–796, 1973.

[5] Hald, O. "Convergence of vortex methods, II," *SIAM J. Numer. Anal.*, Vol. 16, pp. 726-755, 1979.

[6] Richtmyer, R. D. and Morton, K. W. *Difference Methods for Initial Value Problems*, Interscience, New York, 1967.

Chapter 6

Perturbation Methods and Computer Algebra

Richard H. Rand

1   Introduction

The purpose of this paper is to report on some recent work aimed at automating the algebraic computations associated with the approximate solution of systems of nonlinear differential equations by perturbation methods. These methods, which have traditionally been performed by hand, are accomplished more accurately and more efficiently by using computer algebra. The programs which accomplish the following computations have been written in Macsyma [5] and will not be given here, but are available in [6]. We have implemented the following methods on Macsyma (see [6]):

- Linstedt's method

- Center manifold reductions

- Normal forms

- Two variable expansion method (multiple scales)

- Averaging

- Lie transforms for Hamiltonian systems

- Liapunov-Schmidt reductions

In this paper we shall apply the first three methods in the list to examples involving van der Pol's equation:

$$\frac{d^2 x}{dt^2} + x - e(1 - x^2)\frac{dx}{dt} = 0 \tag{1}$$

2 Lindstedt's Method

Van der Pol's equation possesses a periodic solution called a *limit cycle* for all $e \neq 0$ [2]. Lindstedt's method offers a scheme for approximating the limit cycle for small e. If in (1) we set $\tau = \omega t$, expand

$$\begin{aligned}
x(\tau) &= x_0(\tau) + ex_1(\tau) + \cdots + e^n x_n(\tau) & (2) \\
\omega &= \omega_0 + e\omega_1 + \cdots + e^n \omega_n & (3)
\end{aligned}$$

and collect terms, we obtain a sequence of linear ODE's on the x_n's, e.g.,

$$\begin{aligned}
x_0'' + x_0 &= 0 & (4) \\
x_1'' + x_1 &= -2\omega_1 x_0'' + x_0'(1 - x_0^2). & (5)
\end{aligned}$$

Solving these recursively, we obtain at the nth step,

$$x_n'' + x_n = (\cdots)\sin \tau + (\cdots)\cos \tau + \text{ non-resonant terms.} \tag{6}$$

Requiring the coefficients of $\sin \tau$ and $\cos \tau$ to vanish (in order to have a periodic solution) yields values for the amplitude and frequency of the limit cycle to order n. Here are typical results of the Macsyma program ([5],[6]). First we present (2) valid to $O(e^7)$:

```
                        (sin(3 tau) - 3 sin(tau)) e
x(tau) = 2 cos(tau) - ---------------------------
                                  4


                                                    2
        (5 cos(5 tau) - 18 cos(3 tau) + 12 cos(tau)) e
      - ---------------------------------------------
```

$$+ (28 \sin(7\ tau) - 140 \sin(5\ tau) + 189 \sin(3\ tau)$$

$$- 63 \sin(tau))\ e^3 /2304 + (1647 \cos(9\ tau)$$

$$- 10745 \cos(7\ tau) + 21700 \cos(5\ tau)$$

$$- 16920 \cos(3\ tau) + 3285 \cos(tau))\ e^4 /552960$$

$$- (49797 \sin(11\ tau) - 402678 \sin(9\ tau)$$

$$+ 1106210 \sin(7\ tau) - 1322125 \sin(5\ tau)$$

$$+ 582975 \sin(3\ tau) + 194565 \sin(tau))\ e^5 /66355200$$

$$- (10728705 \cos(13\ tau) - 103688277 \cos(11\ tau)$$

$$+ 363275388 \cos(9\ tau) - 595527380 \cos(7\ tau)$$

$$+ 454903750 \cos(5\ tau) - 76288275 \cos(3\ tau)$$

$$- 54423600 \cos(tau))\ e^6 /55738368000$$

$$+ (2340511875 \sin(15\ tau) - 26331793605 \sin(13\ tau)$$

$$+ 112564838232 \sin(11\ tau) - 237436059483 \sin(9\ tau)$$

$$+ 260002792805 \sin(7\ tau) - 118802031250 \sin(5\ tau)$$

$$- 18287608500 \sin(3\ tau) + 34770385650 \sin(tau))\ e^7$$

$$/46820229120000 + \ .\ .\ .$$

Next we display (3) valid to $O(e^8)$:

$$w = 1 - \frac{e^2}{16} + \frac{17\ e^4}{3072} + \frac{35\ e^6}{884736} - \frac{678899\ e^8}{5096079360} + \ .\ .\ .$$

3   Center Manifold Reduction

Center manifold reduction is a technique which simplifies an appropriate problem by focusing attention on nonlinear extensions of critical eigendirections, thereby eliminating unimportant information from consideration. In order to provide an example of this method, we consider the behavior of the van der Pol equation at infinity, i.e., for large values of x and dx/dt [3]. Following Poincare (see [4]) we set in (1)

$$y = \frac{dx}{dt}, \qquad v = \frac{y}{x}, \qquad z = \frac{1}{x}, \qquad d\tau = x^2 dt \qquad (7)$$

to obtain

$$z' = -vz^3$$

$$v' = -ev + z^2(ev - v^2 - 1). \qquad (8)$$

This system of DE's describes the behavior of van der Pol's equation at infinity, which here corresponds to $z = 0$ [3]. An equilibrium exists at the origin, $v = z = 0$. Linearization about the origin shows that the eigenvalues are $-e$ and 0. Corresponding to the 0 eigenvalue is the eigendirection $z = 0$. The center manifold theorem [1] tells us that there exists a (generally curved) line (the center manifold) tangent to this eigendirection which is invariant under the flow (8). The flow along the center manifold determines the stability of the equilibrium point.

In order to obtain an approximate expression for the center manifold, we set

$$z = a_2 v^2 + a_3 v^3 + \cdots \qquad (9)$$

and substitute (9) into (8). Replacing derivatives of z and v by the values prescribed by the flow (8), and collecting terms, we obtain a sequence of algebraic equations on the a_i.

Here are typical results of the Macsyma program [6]. First we present the center manifold to $O(z^{12})$:

```
         2      4      2          6      2          8
        z      z    (e  - 1) z   (e  - 5) z
v = - - -- - -- - ------------ - ------------
        e      e         3                3
                        e                e
```

$$-\frac{(e^4 - 14e^2 + 6)z^{10}}{e^5} - \frac{(e^4 - 30e^2 + 54)z^{12}}{e^5} + \ldots$$

Next we display the flow on the center manifold to $O(z^{15})$:

$$\frac{dz}{dt} = \frac{z^5}{e} + \frac{z^7}{e} + \frac{(e^2 - 1)z^9}{e^3} + \frac{(e^2 - 5)z^{11}}{e^3}$$

$$+ \frac{(e^4 - 14e^2 + 6)z^{13}}{e^5} + \frac{(e^4 - 30e^2 + 54)z^{15}}{e^5} + \ldots$$

In particular, the last result gives that for $e > 0$ the equilibrium $v = z = 0$ is unstable.

4 Normal Forms

Normal forms is a method for obtaining approximate solutions to a system of ordinary differential equations. The method consists of generating a transformation of variables such that the transformed equations in the new variables are in a simplified form. As an example, we again consider the van der Pol equation at infinity. Since the transformation (7) is singular for $x = 0$, the foregoing analysis based on (7) and (8) is invalid near the "ends" of the y-axis. To complete our investigation of the van der Pol equation at infinity, we interchange the roles of x and y in (7). We set in (1)

$$y = \frac{dx}{dt}, \qquad u = \frac{x}{y}, \qquad z = \frac{1}{y}, \qquad d\tau = y^2 dt \qquad (10)$$

which gives

$$z' = uz^3 + ez(u^2 - z^2)$$

$$(11)$$

$$u' = z^2(1 + u^2 - eu) + eu^3.$$

As in (8), $z = 0$ here corresponds to the trajectory at infinity. An equilibrium exists in (11) at $u = z = 0$. We are interested in its stability. Although (11) have no linear part, the substitution

$$w = z^2 \tag{12}$$

produces the system:

$$\begin{aligned} w' &= 2uw^2 + 2ew(u^2 - w) \\[2mm] u' &= w(1 + u^2 - eu) + eu^3 \end{aligned} \tag{13}$$

which has the linear part

$$w' = 0, \qquad u' = w. \tag{14}$$

Although the linearized system (14) has a double zero eigenvalue, and hence is not sufficient to yield conclusions about stability, the question can be settled by the method of normal forms. Takens [9] has shown that systems with this linear part can always be put in the normal form

$$\begin{aligned} p' &= a_2 q^2 + a_3 q^3 + a_4 q^4 + \cdots \\[2mm] q' &= p + b_2 q^2 + b_3 q^3 + b_4 q^4 + \cdots \end{aligned} \tag{15}$$

where p and q are related to w and u via a near-identity transformation.

Here is the result of the Macsyma program which computes the near-identity transformation and the new normal form (see [6],[8]): If the following near-identity transformation is substituted into (13),

```
                          2       2          3
                  (12 e  + 3) q  p + 2 e q
    [w = p - 2 e q p + ---------------------------
                               3

           3          3        2 4
     (24 e  + 15 e) q  p + 7 e  q
   - ---------------------------------
                   3
```

$$+ \frac{(960\,e^4 + 1035\,e^2 + 80)\,q^4 p + (370\,e^3 + 48\,e)\,q^5}{60} + \ldots,$$

$$u = q - \frac{3\,e\,q^2}{2} + \frac{(15\,e^2 + 4)\,q^3}{6} - \frac{(105\,e^3 + 70\,e)\,q^4}{24}$$

$$+ \frac{(945\,e^4 + 1086\,e^2 + 88)\,q^5}{120} + \ldots\,]$$

the following transformed differential equations result:

$$\left[\frac{dp}{dt} = -2\,e^2\,q^5 + \ldots,\right.$$

$$\left.\frac{dq}{dt} = p + \frac{5\,e\,q^3}{3} - \frac{5\,e^2\,q^4}{2} + \frac{(225\,e^3 + 8\,e)\,q^5}{60} + \ldots\,\right]$$

Have we taken enough terms in the series to correctly describe the local behavior of the system in the neighborhood of the origin? A truncated system is said to be *determined* if the inclusion of any higher order terms cannot effect the topological nature of the local behavior near the equilibrium. Using Macsyma, it has been shown [8] that a system of the form:

$$p' = a_5 q^5 + a_6 q^6 + a_7 q^7 + \cdots$$

$$\tag{16}$$

$$q' = p + b_3 q^3 + b_4 q^4 + b_5 q^5 + \cdots$$

is topologically equivalent to the truncated system:

$$p' = a_5 q^5$$

$$\tag{17}$$

$$q' = p + b_3 q^3$$

if

$$a_5 \neq 0 \quad \text{and} \quad 4a_5 + 3b_3^2 > 0. \tag{18}$$

In the foregoing case of van der Pol's equation, (18) are satisfied, and hence the stability of the origin in (11) is the same as in the system:

$$p' = -2e^2 q^5$$

$$q' = p + \frac{5}{3} e q^3. \tag{19}$$

Equations (19) may be simplified by setting $r = q^3$ and reparameterizing time with $dT = q^2 d\tau$ (recall τ was defined in (10)), giving

$$p' = -2e^2 r$$

$$r' = 3p + 5er. \tag{20}$$

Equations (20) are a linear system with eigenvalues $2e$ and $3e$, and thus the origin is unstable for $e > 0$ (and stable for $e < 0$).

A   Appendix

We include in this Appendix the results of a normal form computation for the general two dimensional nilpotent case (of which (13) are a special case):

$$w' = \sum_i \sum_j g_{ij}\, u^i w^j, \qquad u' = w + \sum_i \sum_j f_{ij}\, u^i w^j. \tag{21}$$

in which the summed terms are strictly nonlinear. Equations (21) can be transformed to Takens' normal form (15) by a near-identity transformation, which to $O(3)$ is given by:

```
                   2                            2
          2 k1 p    + 2 g02 q p + g11 q
 [w = p + ---------------------------------
                        2
```

$$+ (6\ k3\ p^3 + (6\ g11\ k2 + 12\ g02\ k1 + 6\ g03)\ q\ p^2$$

$$+ (6\ g20\ k2 + 3\ g11\ f02 + 6\ g11\ k1 + 6\ g02^2 + 3\ g12)$$

$$q^2\ p + (4\ g20\ f02 + g11\ f11 + (-\ 2\ f20 + 4\ g11)\ g02$$

$$+ 2\ g21)\ q^3\ )/6,$$

$$u = q$$

$$+ \frac{2\ k2\ p^2 + (2\ f02 + 2\ k1)\ q\ p + (f11 + g02)\ q^2}{2}$$

$$+ (6\ k4\ p^3 + (6\ f11\ k2 + 12\ k1\ f02 + 6\ f03 + 6\ k3)\ q\ p^2$$

$$+ ((6\ f20 + 3\ g11)\ k2 + (3\ f11 + 6\ g02)\ f02$$

$$+ (6\ f11 + 6\ g02)\ k1 + 3\ f12 + 3\ g03)\ q^2\ p$$

$$+ (-\ 2\ g20\ k2 + (2\ f20 + 2\ g11)\ f02 + (2\ f20 + g11)\ k1$$

$$+ f11^2 + 3\ g02\ f11 + 2\ g02^2 + 2\ f21 + g12)\ q^3\ )/6]$$

The parameters k1, k2, k3, k4 in the above transformation are arbitrary constants resulting from a nonempty null space of the linear equations determining the coefficients of the near-identity transformation [6]. If the near-identity transformation is appropriately extended to terms of $O(5)$, the following normal form may be obtained:

$$[\frac{dp}{dt} = g20\ q^2 + (f11\ g20 - g11\ f20 + g30)\ q^3$$

$$-\ (20\ k2\ g20^2 + ((16\ f02 - 8\ k1)\ f20$$

$$+ (-2\, f02 - 4\, k1)\, g11 - 7\, f11^2 - 6\, g02\, f11 + g02^2$$

$$- 8\, f21 + 2\, g12)\, g20 - 12\, g02\, f20^2$$

$$+ ((6\, f11 + 6\, g02)\, g11 + 12\, g21)\, f20 + 12\, f30\, g11$$

$$- 18\, g30\, f11 - 6\, g02\, g30 - 12\, g40)\, q^4 / 12$$

$$- ((42\, k2\, f11 - 10\, k2\, g02 - 16\, f02^2 + 4\, k1\, f02 - 4\, k1^2$$

$$+ 16\, f03 + 10\, k3)\, g20^2 + ((-42\, k2\, g11 + 12\, f02\, f11$$

$$+ (28\, f02 + 4\, k1)\, g02 + 12\, f12 - 14\, g03)\, f20$$

$$+ ((f02 - 2\, k1)\, f11 + (f02 - 2\, k1)\, g02 - 2\, f12$$

$$- 4\, g03)\, g11 - 3\, f11^3 - 7\, g02\, f11^2$$

$$+ (-3\, g02^2 - 12\, f21 + g12)\, f11 + 42\, k2\, g30 + g02^3$$

$$+ (-8\, f21 + 3\, g12)\, g02 + (-2\, f02 - 4\, k1)\, g21$$

$$+ 12\, f30\, f02 - 12\, f30\, k1 - 6\, f31 + 2\, g22)\, g20$$

$$+ (-12\, f02\, g11 - 12\, g02^2 - 12\, g12)\, f20^2$$

$$+ ((3\, f11^2 + 6\, g02\, f11 + 3\, g02^2)\, g11 + 12\, g21\, f11$$

$$+ 24\, f02\, g30 + (12\, g21 - 24\, f30)\, g02 + 12\, g31)\, f20$$

$$+ (12\, f30\, f11 + 12\, f30\, g02 + 12\, f40)\, g11 - 15\, g30\, f11^2$$

$$+ (-18\, g02\, g30 - 24\, g40)\, f11$$

$$+ (-3\, g02^2 - 12\, f21)\, g30 - 12\, g40\, g02 + 12\, f30\, g21$$

$$- 12\ g50)\ q^5/12 + \ldots,$$

$$\frac{dq}{dt} = p + \frac{(2\ f20 + g11)\ q^2}{2}$$

$$- ((2\ f02 + 6\ k1)\ g20 + 2\ g02\ f20 + (-\ f11 - g02)\ g11$$

$$-\ 2\ g21 - 6\ f30)\ q^3/6 - ((16\ k2\ f20 + 8\ k2\ g11$$

$$+\ 18\ k1\ f11 - 12\ f02\ g02 + 6\ f12 + 18\ g03)\ g20$$

$$+ (8\ f02 + 8\ k1)\ f20^2 + (-\ 10\ k1\ g11 - 2\ f11^2 + 8\ g02^2$$

$$+\ 8\ f21 + 10\ g12)\ f20 + (f02 + 2\ k1)\ g11^2$$

$$+ (-\ f11^2 - 3\ g02\ f11 - 2\ g02^2 - 2\ f21 - g12)\ g11$$

$$+ (-\ 6\ g21 - 12\ f30)\ f11 + (6\ f02 + 18\ k1)\ g30$$

$$+ (-\ 6\ g21 - 6\ f30)\ g02 - 6\ g31 - 24\ f40)\ q^4/24$$

$$+ ((40\ k2\ f02 + 120\ k2\ k1)\ g20^2$$

$$+ ((-\ 60\ k2\ f11 - 100\ k2\ g02 + 48\ f02^2 + 136\ k1\ f02$$

$$-\ 104\ k1^2 + 8\ f03 + 140\ k3)\ f20$$

$$+ (-\ 50\ k2\ f11 - 90\ k2\ g02 + 4\ f02^2 + 8\ k1\ f02$$

$$-\ 52\ k1^2 + 4\ f03 + 70\ k3)\ g11 + (-\ 2\ f02 - 42\ k1)\ f11^2$$

$$+ ((36\ f02 - 36\ k1)\ g02 - 12\ f12 - 72\ g03)\ f11$$

$$+ (50\ f02 + 6\ k1)\ g02^2 - 72\ g03\ g02$$

$$+ (8\, f02 - 48\, k1)\, f21 + (28\, f02 + 12\, k1)\, g12$$

$$- 40\, k2\, g21 - 120\, k2\, f30 - 24\, f22 - 36\, g13)\, g20$$

$$+ (60\, k2\, g11 + (- 32\, f02 - 32\, k1)\, g02 + 44\, g03)\, f20^2$$

$$+ (30\, k2\, g11^2 + ((- 14\, f02 + 16\, k1)\, f11$$

$$+ (- 38\, f02 + 36\, k1)\, g02 + 8\, f12 + 50\, g03)\, g11$$

$$+ 8\, g02\, f11^2 - 22\, g12\, f11 - 60\, k2\, g30 - 32\, g02^3$$

$$+ (- 32\, f21 - 74\, g12)\, g02 + (- 12\, f02 + 32\, k1)\, g21$$

$$- 120\, f30\, f02 - 120\, f30\, k1 - 60\, f31 - 44\, g22)\, f20$$

$$+ ((- 3\, f02 - 10\, k1)\, f11 + (- 3\, f02 - 10\, k1)\, g02$$

$$- 2\, f12 - 4\, g03)\, g11^2 + (f11^3 + 6\, g02\, f11^2$$

$$+ (11\, g02^2 + 8\, f21 + 5\, g12)\, f11 - 30\, k2\, g30 + 6\, g02^3$$

$$+ (12\, f21 + 7\, g12)\, g02 + (- 10\, f02 - 20\, k1)\, g21$$

$$- 12\, f30\, f02 + 12\, f30\, k1 + 6\, f31 + 2\, g22)\, g11$$

$$+ (14\, g21 + 30\, f30)\, f11^2 + ((- 12\, f02 - 108\, k1)\, g30$$

$$+ (36\, g21 + 48\, f30)\, g02 + 36\, g31 + 120\, f40)\, f11$$

$$+ ((36\, f02 - 36\, k1)\, g02 - 24\, f12 - 72\, g03)\, g30$$

$$+ (22\, g21 - 6\, f30)\, g02^2 + (36\, g31 + 96\, f40)\, g02$$

$$+ 16\, g21\, f21 + (8\, g21 - 24\, f30)\, g12$$

$$+ (- 24\, f02 - 72\, k1)\, g40 + 120\, f50 + 24\, g41)\, q^5 /120$$

$$+ \, . \, . \, . \, .]$$

Note that if $g20 \neq 0$, then $k1$ may be chosen so as to kill the q^3 term in the q' equation. Similarly, if $(g11 + 2\ f20)\ g20 \neq 0$, then $k2$ and $k3$ may be respectively chosen to kill the q^4 and q^5 terms in the q' equation.

References

[1] Carr, J., *Applications of Centre Manifold Theory*, Applied Mathematical Sciences, Vol. 35, Springer-Verlag, New York, 1981.

[2] Cesari, L., *Asymptotic Behavior and Stability Problems in Ordinary Differential Equations*, Ergebnisse der Mathematik und ihrer Grenzgebiete, Vol. 16 (Third edition), Springer-Verlag, New York, 1971.

[3] Keith, W. L., and Rand, R. H., "Dynamics of a system exhibiting the global bifurcation of a limit cycle at infinity," *Int. J. Non-Linear Mechanics*, Vol. 20, pp. 325–338, 1985.

[4] Minorsky, N., *Nonlinear Oscillations*, Van Nostrand, Princeton, 1962.

[5] Rand, R. H., *Computer Algebra in Applied Mathematics: An Introduction to Macsyma*, Research Notes in Mathematics, Vol. 94, Pitman Publishing, Boston, 1984.

[6] Rand, R. H., and Armbruster, D., *Perturbation Methods, Bifurcation Theory and Computer Algebra*, Applied Mathematical Sciences, Vol. 65, Springer-Verlag, New York, 1987.

[7] Rand, R. H., and Keith, W. L., "Normal form and center manifold calculations on Macsyma," in *Applications of Computer Algebra*, R. Pavelle, ed., pp. 309–328, Kluwer Academic Publishers, Boston, 1985.

[8] Rand, R. H., and Keith, W. L., "Determinacy of degenerate equilibria with linear part $x' = y$, $y' = 0$ Using Macsyma," *Applied Mathematics and Computation*, Vol. 21, pp. 1–19, 1987.

[9] Takens, F., "Singularities of Vector Fields," *Publ. Math. Inst. Hautes Etudes Sci.*, Vol. 43, pp. 47–100, 1974.

Chapter 7

Multibody Simulation in an Object Oriented Programming Environment

N. Sreenath and P.S. Krishnaprasad

1   Introduction

A multibody system is simply a collection of bodies, rigid or non-rigid, interconnected by means of joints with certain specific kinematic properties [29]. A large class of mechanical systems can be classified as multibody systems. We list here a few examples: spacecraft, robot manipulators, land vehicles (e.g., automobiles), the human body, molecular interconnection of atoms, cables modeled as series of rigid bodies, etc. [30]. A *planar multibody system* is a multibody system with motions of the system restricted to a plane. To study the motions of multibody systems the knowledge of the dynamics of the system is essential. The dynamics of a multibody system cannot be adequately approximated by linear differential equations, since large motions are characteristic of such systems. Further complications arise with the introduction of control variables and external and internal disturbances.

Simulation has been one of the important tools for understanding the motions of a multibody system. The steps involved in simulating the motions of a complex multibody system are as follows. First, one generates a dynamical model of the multibody system in the form of differential equations. Next a simulation program for numerically integrating these differential equations is created. Then the simulation code is run for a relevant set of parameters and a specified initial state of the system to compute the trajectory of the system.

Thus the primary step in understanding the motions of a multibody system is the formulation of the dynamical equations. For the case when the system is in the form of an open kinematic chain (Figure 1), i.e., the path from one body in the system to any other body in this system is unique, and each body in the system can be modeled as a rigid body, we get a set of coupled, highly non-linear first or second-order ordinary differential equations.

Multibody dynamical formalism has been the subject of a lot of research [6], [9], [10], [14],[16], [20], [22], [23], [32], [28], [29]. The formulation of dynamical equations by hand is a tedious process and often prone to errors. Many researchers have considered the possibility of computer-aided methods to generate these equations. General purpose multibody computer programs capable of generating the dynamical equations as well as simulating them are available for quite sometime (see [21], [23] for references).

These computer-oriented methods may be classified as numerical [7] and symbolic programs [3], [5], [15], [18], [21], [24], [27], [29]. Numerical programs are characterized by numerical digital computation whereas symbolic programs generate equations and accompanying expressions in symbolic (or alpha-numeric) form on the basis of alpha-numeric data. Symbolic programs in general are more efficient in terms of running time in comparison with numerical programs [23], [31].

In this paper we present a general-purpose software system architecture designed to generate the dynamical equations of a multibody system *symbolically* [1], automatically generate the computer code to simulate these equations *numerically*, run the simulation, and display the results graphically. This architecture is implemented for planar two- and three-body systems in the package called OOPSS - Object Oriented Planar System Simulator. A nice *user interface* is part of OOPSS.

The paper is organized as follows. In Section 2 we discuss the system features, followed by a brief exposition on multibody dynamics in Section 3. The Object Oriented Programming methodology

---

[1]Only multibody systems connected in the form of a tree (Figure 1) with pin joints are considered here.

and the software architecture of OOPSS are discussed in Section 4. Section 5 discusses the implementation of OOPSS architecture for planar two- and three-body systems followed by the conclusion in Section 6. The Appendix contains the derivation of the reduced Hamiltonian and the dynamics in terms of the Poisson bracket for a planar multibody system, along with a planar two-body example.

2   System Features

OOPSS uses Object Oriented Programming (OOP) along with *symbolic manipulation* to formulate and simulate the dynamics of a planar multibody system automatically. A mathematical model describing the motion of a planar multibody system (dynamic model) is generated by OOPSS symbolically. The symbolic manipulation has been implemented in Macsyma[2]. A program to numerically simulate these differential equations is generated. OOPSS animates the multibody system by exploiting the high-resolution graphics and windowing facilities of a Lisp machine and has been implemented in Zeta-Lisp on a Symbolics 3600[3] series machine. A nice *user interface* is provided for interacting with the symbolic, numeric, and graphic elements of OOPSS.

Users can interactively:

1. choose any kinematic or physical parameters for the system,

2. change any runtime initial condition—system energy, system angular momentum, time step, maximum problem time, initial values of state, and other variables (angles, conjugate momentum variables),

3. select display parameters for the graphs,

4. choose feedback control torque laws and gains.

One of the significant features of OOPSS is that various control schemes can be easily implemented and evaluated. Currently *proportional, proportional-derivative, sinusoidal-spring*, and *sinusoidal-spring with bias* feedback control schemes have been implemented.

---

[2] Macsyma is a trademark of Symbolics Inc., Cambridge, Mass.

[3] Manufactured by Symbolics Inc., Cambridge, Mass.

These control laws can be interactively selected, evaluated, and the control gains tuned. A model-dependent control scheme, for example, exact linearization [11], [12], could be easily implemented since the associated feedback control law could be formulated using symbolic computation.

OOPSS can be used a design tool to design the multibody system parameters. It can be used as an experimental work-bench to study certain problems of mechanics. To enhance our knowledge about the phase space of a multibody system, we could simulate the equilibria of the multibody system and explore the stability of such equilibria.

3   Multibody Motion

Before embarking on the description of the OOPSS system it is necessary to introduce the various vectors, parameters, and variables associated with the motion of a multibody system. Associated with every body in the system are physical parameters such as mass, inertia, and various kinematic parameters. The vectors connecting the joint and the center of mass of the body, and the same joint and corresponding joints on an adjacent body, provide information on the kinematic description of the multibody system. Since the individual bodies are considered rigid at any time instant, the orientation of the body and the location of its center of mass provide sufficient information to determine the configuration of the body. Other quantities such as angular velocity and angular acceleration are associated with each body, and quantities such as kinetic energy, angular momentum, and linear momentum are associated with the multibody system itself [4].

The motions of the multibody system are defined with respect to an inertial coordinate system. For a planar multibody system, a local coordinate system is defined for every body in the system, and an angle this coordinate system makes with respect to the inertial

---

[4]To define the multibody system mathematically it is necessary to label every body and every joint in the system uniquely in increasing order of consecutive integers. Also for simplicity of computation it is convenient to label the bodies such that the body labels are of increasing magnitude along any topological path starting at body 1; the joint connecting a body i to the body with a lesser body-label is labelled as joint $(i\text{-}1)$. The joint $(i\text{-}1)$ is also known as the *previous joint* of body i.

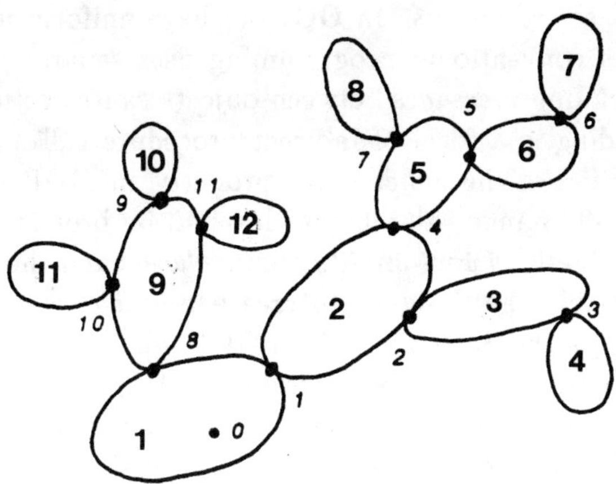

Figure 1. Multibody system connected in the form of a tree.

coordinate system is the inertial angle associated with the body.
The angle between the local coordinate systems of two bodies is the
relative angle between the two bodies.

The dynamics of a planar multibody system connected in the
form of a tree can be described by a set of first-order differential
equations in terms of the Hamiltonian of the system H^5 (which is
also the kinetic energy of the system) [23], the relative angles $\theta_{i,j}$
between the bodies, and the conjugate momenta μ_i.

4   System Description

OOPSS is implemented using the OOP technique. We start with a
brief introduction to the OOP methodology and discuss the important **flavors, methods,** and **functions** used in the program. The
system architecture is discussed in the next section.

4.1   Flavors, Methods, and Functions

Objects are entities that combine the properties of procedures and
data, since they perform computation and save local state [26]. Also,
objects could be linked to real world things. A program could be

---

[5] See the Appendix for a planar two-body system dynamics example.

built using a set of *objects*. In OOP we have uniform usage of objects whereas conventional programming uses separate procedures and data. Sending *messages* between objects causes action in OOP. Message sending is a form of indirect procedure call and supports *data abstraction*. The *inheritance* property in OOP enables the transmission of changes at a higher level to be broadcast throughout the lower levels. *Functionality encapsulation* and the *inheritance* properties enable the designer to create *reusable software* components termed as the *Software-IC*'s in OOP [4].

We have used *Zeta-Lisp*, a programming language capable of object oriented programming. General descriptions of objects in *Zeta-Lisp* are in the form of **flavors**. **Flavors** are abstract types of object class. In Zeta-Lisp a conceptual class of objects and their operations are realized by the **Flavor System**, where part of its implementation is simply a convention in procedure calling style; part is a powerful language feature, called **Flavors**, for defining classes of abstract objects. Flavors have *inheritance* property; thus if we build a flavor using other flavors then all the properties of the latter are inherited by the former. Any particular object is an *instance* of a flavor. The variables associated with a generic object are known as *instance variables*.

Specific operations could be associated with the objects using *methods*. One can create a method to define a specific operation on any instance of a flavor and attribute special properties to it. For instance, one can define a method for a body which is the leaf of a tree and so has only one joint (i.e., only one body attached to it—contiguous to and inboard) whereas a generic body has two or more bodies attached to it. Functions are used to send messages to instances of flavors through the already defined methods.

The OOP methodology is conducive to building *iconic user* interfaces since all except the necessary information can be hidden so as to limit clutter and enhance clarity of the issues under consideration. *Methods* provide the basic abstraction of a class of the objects via the *inheritance* property of the OOP.

In the planar multibody setting the primary flavor used to describe a *generic body* in the system is:

general-planar-body.

A *generic body* in the planar multibody system can be defined as an object with the following instance variables: a vector connecting the previous joint to the center of mass of body, vector(s) connecting the previous joint and other joint(s) on the body, and the angle made by the body frame with respect to the inertial coordinate system (also called orientation of the body).

The center of mass of a *generic body* is defined by the use of the method **:center-of-mass**; this ensures that at any time instance knowing the location of the previous joint and the orientation of the body, the body center of mass could be calculated. The information about the orientation of the body is calculated in the **Numerical Simulator** using the dynamical equations. The position of the next joint is defined by using the method **:next-joint**. Note here that the next joint of one body will be the previous joint of another body.

Similarly the **:draw-body** method is used to draw the *generic body* for animation on the screen. This method utilizes the center of mass, the next joint information, and the orientation of the body to create the animation of the multibody motion.

Another flavor

graphics-window-frame

is implemented to create the *graphics window* and the various *panes* associated with it, for animation and data display purposes. This flavor also implements the *menu* and mouse selectable capabilities for the OOPSS system. A number of *functions* are implemented to send messages to the flavors; a few important functions are listed and their functions are self-explanatory: *set-parameters, set-window-frame, on-line-info, show-herald*, etc.

4.2   System Architecture

The OOPSS system can be represented by: **Symbolic Equation Generator, Numerical Simulator, Descriptor,** and **Display** modules which are interconnected as shown in Figure 2. A detailed description of each module is given in the following paragraphs.

4.2.1   Symbolic Equation Generator

The symbolic equation generator generates the dynamics of a planar multibody system connected in the form of a tree structure in the

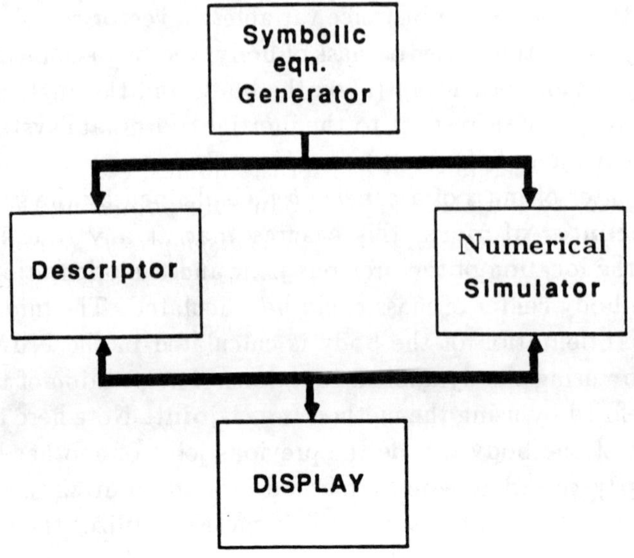

Figure 2. Block diagram representation of OOPSS.

Hamiltonian setting (see Sreenath [23] Chapter 3, for the formulation and notation). This module is implemented in Macsyma. The input data for this module consists of information describing the way the bodies are interconnected, kinematic and physical parameters like lengths, mass, inertia, etc., any control (actuating) or disturbance torques acting on the multibody system. For a two-body system example one form of the output equations from this module are shown in the Appendix.

4.2.2   Numerical Simulator

The **Numerical Simulator,** which simulates the dynamical equations generated by the symbolic equation generator, has been implemented using Fortran-77 running on the Lisp machine. The **Numerical Simulator** needs numerical values of all parameters to be in a data file. This input data file is generated by the **DESCRIPTOR.** The file contains the numerical values of all kinematic and physical parameters, the system angular momentum and system energy values, problem time, time step, etc., associated with the particular example.

The state and related variables (for example, angular velocities) at any instance of time could be passed onto the **DISPLAY** module by means of *lispfunctions* to be used for animation and display purposes. The *lispfunction – displaybody* passes the relevant variables like orientation, angular velocities of the bodies, etc., from the Fortran program to the **DISPLAY** module for animating the motion of the multibody system. The function '*displaybody*' is implemented in Zeta-Lisp in the **DISPLAY** package for the actual animation.

The initial condition for initiating the simulation is chosen such that the the various physical laws governing the conservation of select quantities (such as system angular momentum, system energy) are satisfied.

4.2.3   DESCRIPTOR

The **DESCRIPTOR** consists of descriptions of various flavors to implement the display and the user interface. It also contains flavors to define a generic body in the multibody system. Using *methods* we can attribute special properties to the instances of these flavors. This module also functions as an intermediary between the user interface and the **Numerical Simulator** module by generating the input data file for the Fortran program before a numeric simulation run is started, based on input from the user.

4.2.4   DISPLAY

The **DISPLAY** module is the implementation of various functions to drive the instances of flavors by sending messages to them. **DISPLAY** keeps track of sending proper messages to the relevant panes as and when it receives data from the **Numerical Simulator** module. **DISPLAY** is characterized by a '*tv:alu-xor*' option which helps in erasing the display at time t and creating a new display at time $t + \Delta t$. '*displaybody*' and '*cleanbody*' are functions to display the system and clean the displayed picture off the relevant pane respectively.

The *user interface* consists of a window with many panes (Figure 3). Three frames of references in the corresponding window panes: inertial frame of reference and two other selectable frames (from various joint and/or body frames), have been implemented.

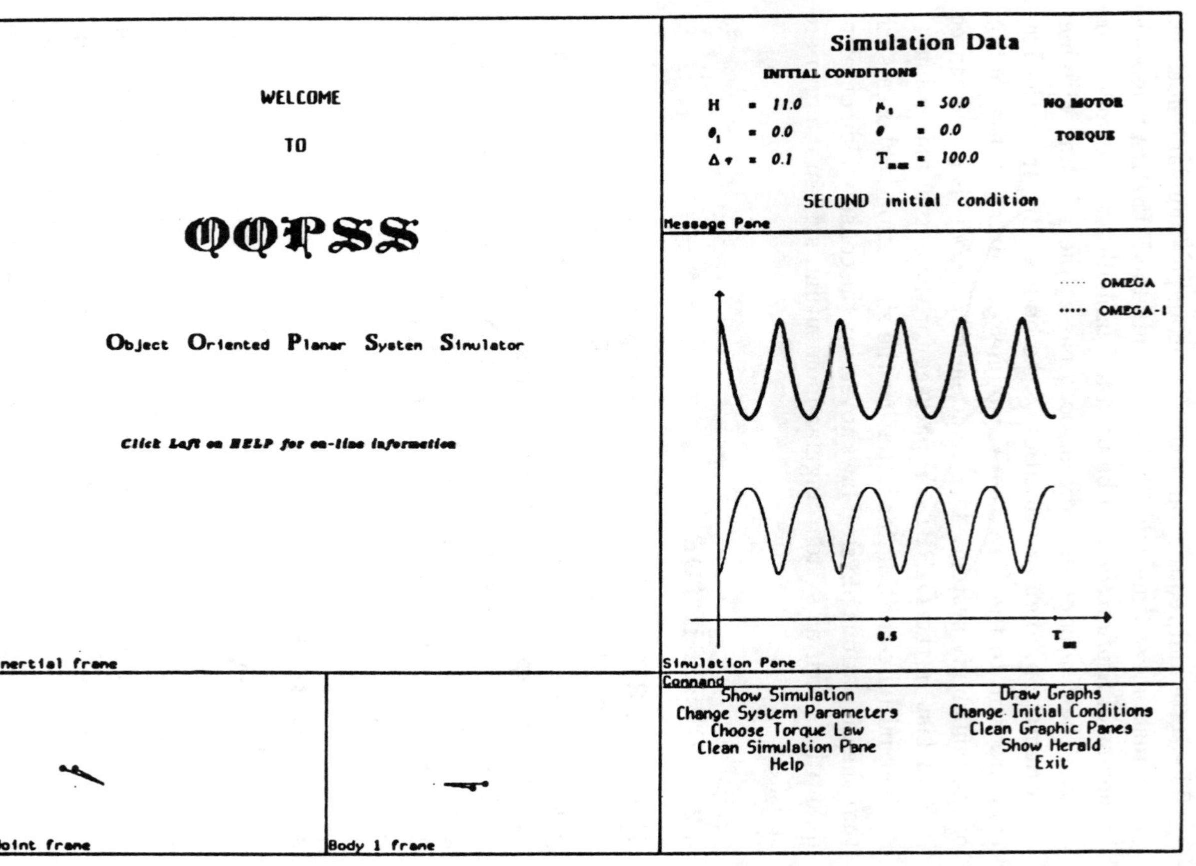

Figure 3. OOPSS window.

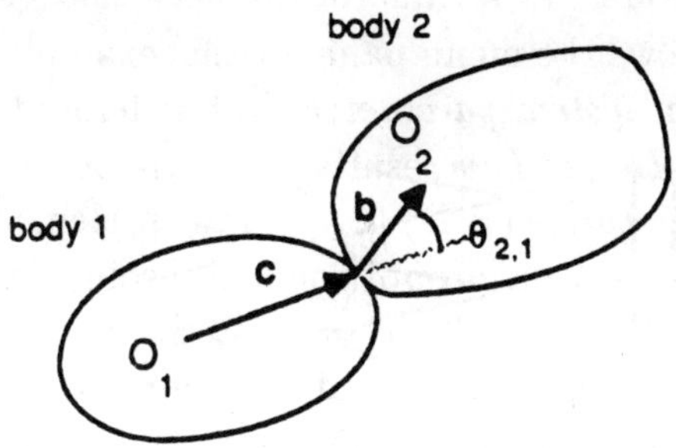

Figure 4. Planar two-body system.

The animation of the multibody system in the selected frames of reference are displayed in their respective panes. A part of runtime data from the Fortran program is displayed in the *message pane*. Graphs of state and/or other variables are drawn as functions of time in the *simulation pane*. The *menu pane* which is in the lower right-hand corner is self-descriptive. Every item in this pane is *mouse sensitive* (mouse selectable). We will say more on this in the following sections as we deal with particular examples of planar two-body and planar three-body systems. A brief online *HELP* facility exists and information can be accessed by clicking left using the mouse when it is highlighted.

5   Implementation

Presently the OOPSS system architecture has been implemented for the planar two-body and the planar three-body systems. The following examples illustrate the capabilities of the OOPSS system.

5.1   Two-Body System

A system of two bodies in space connected together by a one degree of freedom joint occurs in many contexts (Figure 4). One of the bodies, for example, may be a large space based sensor (say a space telescope). Rapid re-orientation of the sensor from one stationary position to another may be desirable. Such a "step-stare" maneuver would require a through knowledge of the dynamics and thus a good mathematical model[6] to formulate the necessary control [19].

Figure 5a shows the menu pane for this example. Clicking left on the mouse when *system parameters* is highlighted gives Figure 5b. Clicking left on *Torque Law* results in Figure 5c. Further clicking on *Proportional-Derivative* (P-D) torque law implements a joint torque law (internal torque)—a proportional sinusoidal biased spring plus derivative controller, i.e., $T_{\text{joint}} = (K_p \sin(\theta_{2,1} - \theta_{\text{bias}}) + K_d \dot{\theta}_{2,1})$. Gains K_p (proportional gain) and K_d (derivative gain), $\theta_{2,1}$ is the relative angle between body 1 and body 2, and, the *bias* angle θ_{bias} could be chosen (see Figure 5d) interactively by the user.

The evolution of motion of a two-body system with a fixed value of system energy and angular momentum is given in Figure 6. The largest *pane* of the display window is the *Inertial frame* where the motion of the system could be observed in the inertial coordinate frame. Each body is represented by a stick figure with a big *filled-in circle* at one end (representing the center of mass of the body) and a smaller circle at the other end representing the joint connecting the other body. The center of mass of the multibody system is defined by the point in the center of the *inertial* frame. Since no external force is present on the system, the center of mass of the system is a fixed point in inertial frame. Two other coordinate frames where the motion of the system could be observed—the *joint frame* and the *Body-1 frame*[7] are also displayed below the *inertial frame*.

The two-body system has two relative equilibria[8] when the bodies are in a extended position (stable) and when the bodies are in folded position (unstable). There is a *homoclinic orbit* associated with the unstable equilibrium point. The *stable equilibrium* is displayed in Figure 7. The trace shown in the inertial frame is the *trace* left by the joint as it moves in space. Simple calculation shows that this trace is indeed a circle when the system is in stable equilibrium position. Figure 8 displays a trajectory when the system

---

[6]Refer to Sreenath [23], Chapter 4 for a detailed description of the problem.

[7]The *joint-frame* is a frame parallel to the inertial coordinate frame with the origin at the joint; the *Body-1 frame* is the local coordinate frame of body 1 located at the body center of mass.

[8]When all the bodies in the system are rotating with constant inertial angular velocity and no relative motion we have relative equilibrium

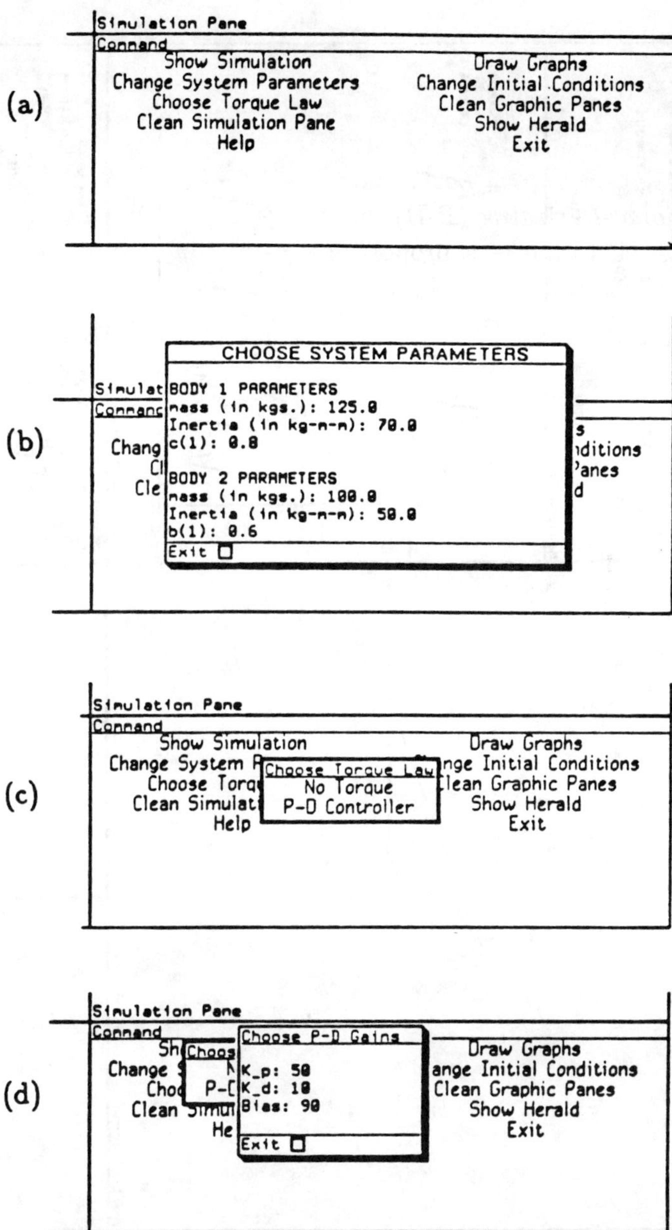

Figure 5. Two-body problem menu-pane.

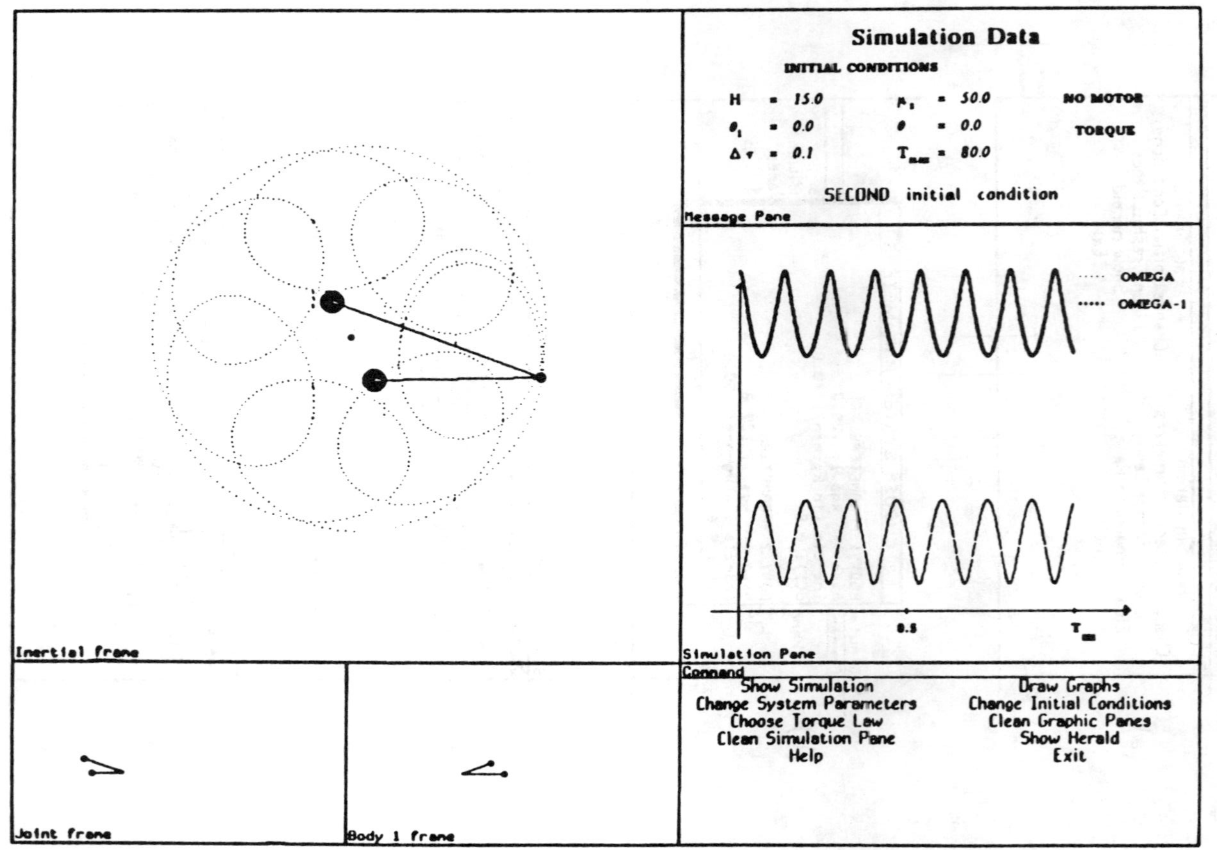

Figure 6. Two-body problem: $H = 15$.

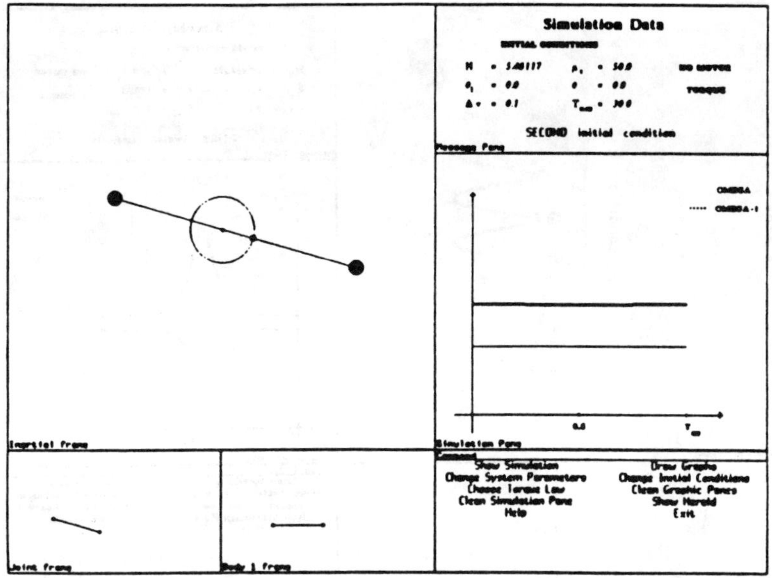

Figure 7. Two-body problem: stable equilibrium.

is at a point very near the *unstable equilibrium* point. If a (P-D) torque is introduced in the system then the resulting trajectory is as shown in Figure 9 (no bias) and Figure 10 (bias). Notice that with K_p equal to zero and K_d positive (Figures 9-10) the system always goes to the stable equilibrium and confirms the 'stabilization theorem'—Theorem 6.3.1 in Sreenath [23], i.e., introduction of a feedback internal torque proportional to the rate of the relative angle stabilizes the system. One could also interpret this result as follows: by introducing this torque law the energy in the system is dissipated till the system goes to a minimum energy state which is the stable (stretched out) relative equilibrium.

5.2   Three-Body System

A general three-body example (Figure 11) has also been implemented on the basis of the OOPSS architecture. Figure 12 shows a general three-body system wherein the center of mass of the middle body is not along the line joining the two joints. The *filled-in circles* represents the center of mass of each body (the first body represented

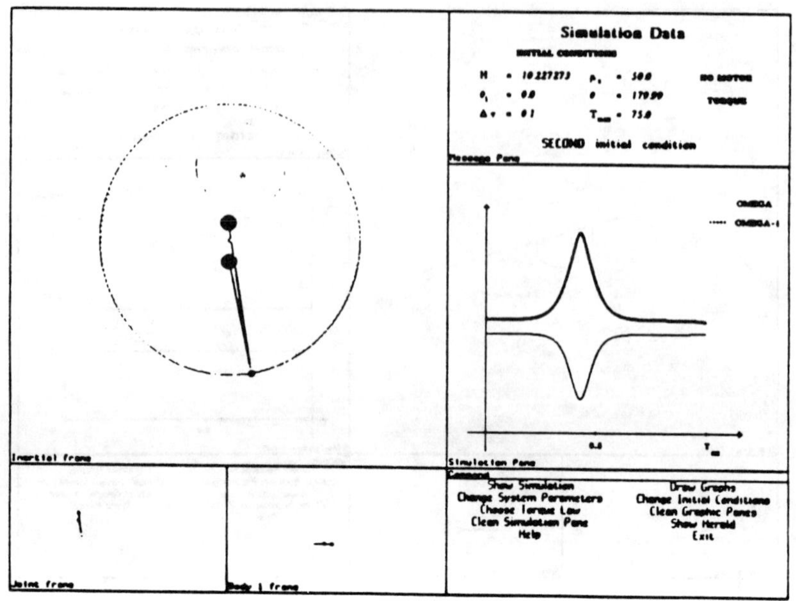

Figure 8. Two-body problem: unstable equilibrium.

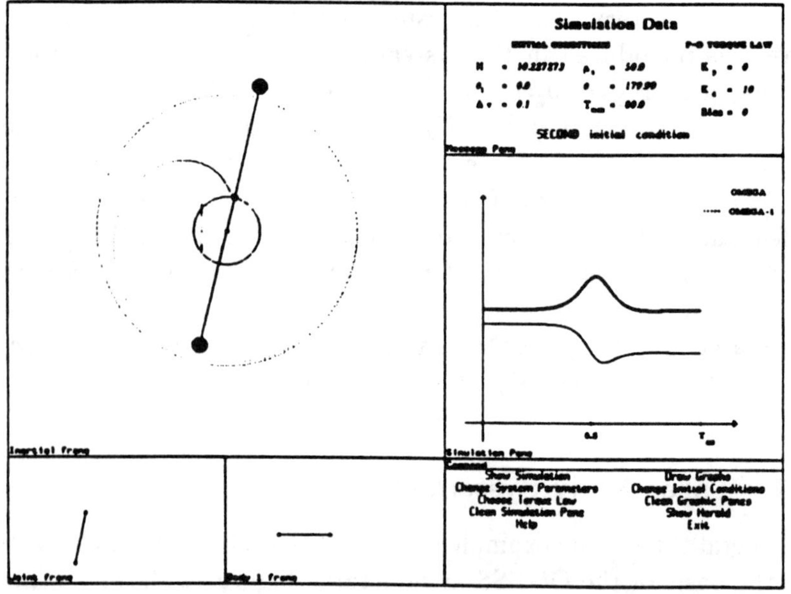

Figure 9. Two-body problem: $K_d = 10$.

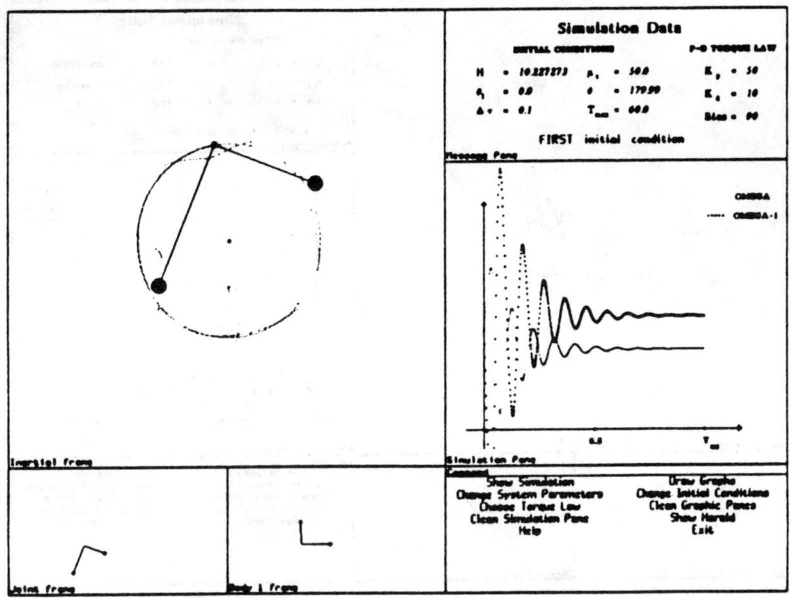

Figure 10. Two-body problem: $K_p = 50$, $K_d = 10$, bias=90.

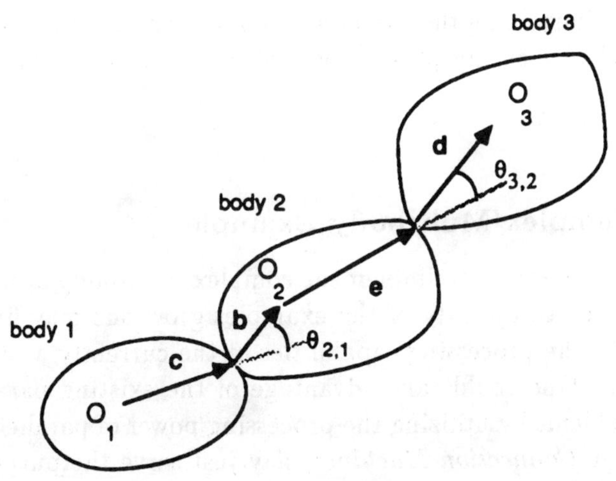

Figure 11. Planar three-body system.

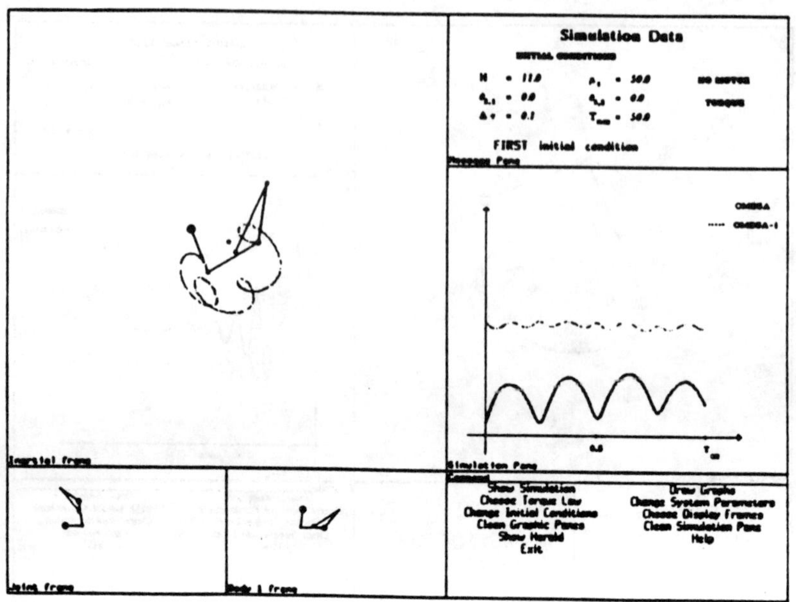

Figure 12. Three-body problem: general case.

by a big circle, second with a smaller circle, and the third with the smallest circle). Display frames could be chosen by clicking left on the 'Choose display frames' using the mouse. Figure 13 displays special kinematic case where the center of mass of the middle body is along the line joining the two joints. Joint torques of the proportional sinusoidal bias spring plus derivative type have been introduced at the joints.

5.3   Complex Multibody Examples

Plans are underway to implement complex multibody system examples. As the complexity of the examples grow one may find oneself limited by the processing capabilities of the currently available Lisp machines. One could take advantage of the existing parallelism in these problems by utilizing the processing power of parallel Lisp processors. A *Connection Machine*[9] may just serve the purpose. Thus dynamics of complex multibody systems may be generated automatically, simulated and animated.

---

[9]Manufactured by Thinking Machines, Inc., Cambridge, Mass.

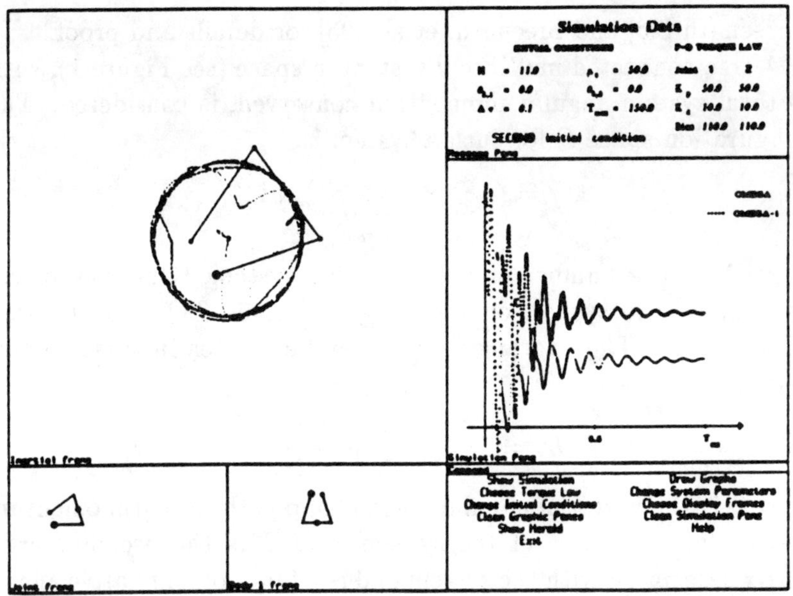

Figure 13. Three-body problem: special kinematic case with joint torques.

6   Conclusion

A prototype object oriented software architecture for multibody system simulation is discussed. OOPSS can generate the dynamic model of a planar multibody system symbolically, generate the computer code to simulate it numerically, run the simulation, and display the result by means of animation and graphs. OOPSS could be used as a test-bed to evaluate control algorithms, select control gains, and design the system parameters for the multibody system, interactively. The system has been successfully implemented for planar two- and three-body systems on a Symbolics 3600 series machine in Zeta-Lisp, Fortran-77 and Macsyma. Plans are underway to implement the OOPSS architecture for more complex multibody systems.

A   Appendix

We develop the dynamical equations of a planar multibody system in space in terms of the Poisson bracket. The symmetries (translational and rotational) acting on the system are taken into account to reduce the dynamics appropriately. We refer the interested reader to Sreenath [23] and Sreenath, et al. [25] for details and proof.

A tree-connected multibody system in space (see Figure 1), with the total system angular momentum conserved, is considered. The configuration space Q for such a system is

$$Q \cong (\underbrace{S^1 \times \cdots \times S^1}_{N \ times}) \times \mathbb{R}^2,$$

where N is the number of bodies in the system. One way of coordinatizing the system on the tangent bundle TQ is by (θ_i, ω_i), $i = 1, \ldots, N$. The Lagrangian can then be written in these coordinates as

$$L = \frac{1}{2}\underline{\omega}^T \mathbf{J}\underline{\omega} + \frac{\| \mathbf{p} \|^2}{2m},$$

where $\underline{\omega}$ is the vector of angular velocities, $\mathbf{p}$ is the linear momentum of the center of mass of the system, and $\mathbf{J}$ is the pseudo-inertia matrix associated with the system and is a function of relative angles between the bodies[10].

The Hamiltonian is simply the kinetic energy of the system and can be constructed using the Legendre transformation as

$$H = \frac{1}{2}\underline{\mu}^T \mathbf{J}^{-1}\underline{\mu} + \frac{\| \mathbf{p} \|^2}{2\mathbf{m}},$$

where $\underline{\mu}$ is the conjugate momentum vector and is related to $\underline{\omega}$ by

$$\underline{\mu} = \mathbf{J}\underline{\omega}.$$

We now recognize the symmetries in the system and reduce the dynamics accordingly. The reduction technique we use is originally due to Arnold [1] and developed further by Marsden and Weinstein [17]. In the general setting one starts with a Poisson manifold P and a Lie group G acting on P by canonical transformations. The reduced phase space P/G (provided it has no singularities) has a natural Poisson structure whose symplectic leaves are the Marsden-Weinstein-Meyer spaces $J^{-1}(\mu)/G_\mu \approx J^{-1}(\mathcal{O})/G$ where $\mu \in \mathfrak{g}^*$, the dual of the Lie algebra of G, $J : P \to \mathfrak{g}^*$ is an equivariant momentum map for the action of G on P, G_μ is the isotropy group of μ (relative to the coadjoint action), i.e., $G_\mu = \left\{ g \in G : Ad^*_{g^{-1}}\mu = \mu \right\}$, and $\mathcal{O}$ is the coadjoint orbit through μ. The coadjoint orbit $\mathcal{O}$, is even dimensional.

---

[10]For example, for the case of planar two-body system $\mathbf{J} = \mathbf{J}(\theta_{2,1})$.

A.1 Reduction by Translations
(Planar Multibody Problems)

We reduce the dynamics by the action of the translation group $\mathbb{R}^2$. This group acts on the original configuration space Q by

$$\mathbf{v} \cdot ((R(\theta_1), \mathbf{r}_1), \ldots, (R(\theta_N), \mathbf{r}_N))$$

$$= ((R(\theta_1), \mathbf{r}_1 + \mathbf{v}), \ldots, (R(\theta_N), \mathbf{r}_N + \mathbf{v}))$$

where $\mathbf{r}_i$, $i = 1, \ldots, N$ is the vector from the origin of the frame of reference to the center of mass of body i; θ_i is the angle made by the body i with respect to the frame of reference. $R(\theta_i)$ is the (2×2) rotation matrix associated with body i,

$$R(\theta_i) \;=\; \begin{bmatrix} \cos(\theta_i) & -\sin(\theta_i) \\ \sin(\theta_i) & \cos(\theta_i) \end{bmatrix}, \qquad i = 1, \ldots, N.$$

The induced momentum map on TQ is calculated by the standard formula

$$J_\xi \;=\; \frac{\partial L}{\partial \dot{q}_i} \xi_Q^i(q),$$

or on T^*Q by

$$J_\xi \;=\; p_i \xi_Q^i(q),$$

where ξ^i is the infinitesimal generator of the action on Q. (see Abraham and Marsden [2]). Coordinatizing Q by θ_i, $i = 1, \ldots, N$ we determine,

$$J_\xi \;=\; \langle \mathbf{p}, \xi \rangle, \; \xi \in \mathbb{R}^2.$$

Thus $J = \mathbf{p}$ is conserved since H is cyclic in $\mathbf{r}$ and so is translation invariant. The corresponding reduced space is obtained by fixing $\mathbf{p} = \mathbf{p}_0$ and letting

$$P_{\mathbf{p}_0} \;=\; J^{-1}(\mathbf{p}_0)/\mathbb{R}^2,$$

(see Chapter 4, Abraham and Marsden [2]). But P_{p_0} is clearly isomorphic to

$$T^*(\underbrace{S^1 \times \cdots \times S^1}_{N \; times})$$

i.e., to the space of $\theta_1, \ldots, \theta_N$ and their conjugate momenta (μ_1, $\ldots$, μ_N). The reduced Hamiltonian is simply the Hamiltonian as before with $\mathbf{p}$ regarded as a constant.

We can adjust H by a constant and thus assume $\mathbf{p} = 0$; this obviously does not affect the equations of motion. Thus,

$$H \;=\; \frac{1}{2}\underline{\mu}^T \mathbf{J}^{-1} \underline{\mu}.$$

Furthermore, the configuration space Q after reduction by translations i.e., reduction to the center of mass frame is

$$Q \;\cong\; \underbrace{(S^1 \times \cdots \times S^1}_{N \ times}).$$

A.2   Reduction by Rotations

The rotational symmetry group S^1 acts on the cotangent space as below:

$$
\begin{aligned}
\theta \cdot ((\theta_1,\mu_1), \cdots, (\theta_N,\mu_N)) &= ((\theta_1 + \theta,\mu_1), \cdots, (\theta_N + \theta,\mu_N)) \\
&= ((\theta_1,\mu_1), \cdots, (\theta_N,\mu_N)).
\end{aligned}
$$

Since S^1 is diffeomorphic to $SO(2)$ (the special orthogonal group of (2×2) matrices) and the Lie algebra of $SO(2)$ is $so(2)$ (skew-symmetric matrices with determinant not equal to zero), the momentum map can be viewed as a map

$$J \;\; : \;\; T^*Q \to so^*(2)$$

where $so^*(2)$ is the dual of the Lie algebra of $SO(2)$

Let $\xi \in so(2)$, then $\exp(t\xi) \in O(2)$. The *infinitesimal generator* $\xi_Q(q)$ can now be calculated as follows:

$$
\begin{aligned}
\xi_Q(q) &= \frac{d}{dt}\Phi(\exp(t\xi),q)\,|_{t-0} \\
&= \frac{d}{dt}(\theta_1 + t, \ldots, \theta_N + t)\,|_{t=0} \\
&= (1, \ldots, 1).
\end{aligned}
$$

The *momentum map* is given by

$$J : T^*Q \;\; \to \;\; so^*(2),$$

with the *momentum* $\mathbf{P} : TQ \to \mathbb{R}$

$$
\begin{aligned}
P(v_q) &= \hat{J}(\xi)(v_q) \\
&= FL(v_q) \cdot \xi_Q(q) \\
&= \langle v_Q, \xi_Q(q) \rangle \quad \text{on } TQ.
\end{aligned}
$$

where $FL : TQ \to T^*Q$ is the *fiber derivative*.

The metric here is the Riemannian metric associated to kinetic energy and the inner product '$\langle , \rangle$' is given by $\langle \underline{x}, \underline{y} \rangle = \underline{x}^T \mathbf{J} \underline{y}$.

$$
\begin{aligned}
\mathbf{P}(v_q) &= \langle (\omega_1, \ldots, \omega_N) , (1, \cdots, 1) \rangle \\
&= [1, \ldots, 1] \mathbf{J} \underline{\omega} \quad \text{on } TQ,
\end{aligned}
$$

or on T^*Q, for $\alpha_q \in T^*Q$ we have

$$
\mathbf{P}(\alpha_q) = \mu_1 + \cdots + \mu_N,
$$

i.e.,

$$
J((\theta_1, \mu_1), \ldots, (\theta_N, \mu_N)) = \mu_1 + \cdots + \mu_N.
$$

We now form the Poisson reduced space

$$
P := T^*(\underbrace{S^1 \times \cdots \times S^1}_{N \; times})/S^1
$$

whose symplectic leaves are the reduced symplectic manifolds

$$
P_\mu = J^{-1}(\mu)/S^1 \subset P.
$$

We coordinatize P by $\theta_{k,J(k)} = \theta_k - \theta_{J(k)}$, $k = 2, \ldots, N$, and μ_j, $j = 1, \ldots N$, where $J(k)$ is the body label of the body connected to body k and $J(k) < k$ via the *previous joint* $(i-1)$ (in Figure 1, $J(5) = 2$, , $J(3) = 2$ and $J(2) = 1$, etc.; also see Footnote 4).

Topologically,

$$
P = \underbrace{S^1 \times \cdots \times S^1}_{(N-1) \; times} \times \mathbb{R}^N.
$$

The Poisson structure on P is computed in the standard way: take two functions

$$
F(\theta_{k,J(k)}, k = 1, \ldots, N, \mu_1, \ldots, \mu_N)
$$

and

$$H(\theta_{k,J(k)}, k = 2,\ldots,N,\ \mu_1,\ldots,\mu_N).$$

Regard them as functions of $\theta_1,\ldots,\theta_N,\mu_1,\ldots,\mu_N$ by substituting $\theta_{k,J(k)} = \theta_k - \theta_{J(k)}$ and compute the canonical bracket.

We can now state our main theorem on the equations of motion of planar multibody systems connected in the form of a tree in terms of the non-canonical bracket.

Theorem A.1 *The dynamics of a multibody system evolve over a reduced Poisson space P coordinatized by $\theta_{k,J(k)}$, $k = 2,\ldots,N$ and μ_k, $k = 1,\ldots N$. Topologically P is $\underbrace{S^1 \times \cdots \times S^1}_{N-1\ times} \times \mathbb{R}^N$. The system is Hamiltonian in the Poisson structure of P with the non-canonical bracket $\{f,g\}$ given by:*

$$\sum_{k=2}^{N}\left[\left(\frac{\partial f}{\partial \mu_{J(k)}} - \frac{\partial f}{\partial \mu_k}\right)\frac{\partial g}{\partial \theta_{k,J(k)}} - \left(\frac{\partial g}{\partial \mu_{J(k)}} - \frac{\partial g}{\partial \mu_k}\right)\frac{\partial f}{\partial \theta_{k,J(k)}}\right],$$

where $f,g : \mathbb{R}^{2N-1} \to \mathbb{R}$.

The corresponding dynamics in terms of the bracket are

$$\begin{aligned}
\dot{\mu}_k &= \{\mu_k, H\} & k = 1,\cdots,N,\\
\dot{\theta}_{k,J(k)} &= \left\{\theta_{k,J(k)}, H\right\} & k = 2,\cdots,N.
\end{aligned}$$

Proof. See Sreenath [23], Chapter 3.

Corollary A.2 *The sum of all the conjugate momentum variables μ_k, $k = 1,\ldots,N$ is equal to the angular momentum of the multibody system, in the center of mass frame.*

Proof. See Sreenath [23], Chapter 3.

A.3   Planar Two-Body System

The dynamics of a planar two-body system with a torque T_{joint} acting at the joint, is given by the following equations:

$$\dot{\mu}_1 = \frac{\partial H}{\partial \theta_{2,1}} + T_{\text{joint}},$$

$$\dot{\mu}_2 = -\frac{\partial H}{\partial \theta_{2,1}} - T_{\text{joint}},$$

$$\dot{\theta}_{2,1} = \frac{\partial H}{\partial \mu_2} - \frac{\partial H}{\partial \mu_1}.$$

The Hamiltonian (kinetic energy) of the planar two-body problem is given by

$$H = \frac{1}{2}[\mu_1, \mu_2]\mathbf{J}^{-1}[\mu_1, \mu_2]^T$$

where the $\mathbf{J}$ is a symmetric pseudo-inertia matrix dependent on the relative angle $\theta_{2,1}$ between the bodies. μ_1 and μ_2 are the conjugate momentum variables and are related to the angular velocities ω_1 and ω_2 of the bodies as below:

$$\begin{bmatrix} \mu_1 \\ \mu_2 \end{bmatrix} = \mathbf{J} \begin{bmatrix} \omega_1 \\ \omega_2 \end{bmatrix}.$$

Also,

$$\mu_1 + \mu_2 = \mu_s$$

where μ_s is the angular momentum of the system, a conserved quantity.

References

[1] Arnold, V. I., *Mathematical Methods of Classical Mechanics*, Springer Verlag, New York, 1978.

[2] Abraham, R., and Marsden, J. E., *Foundations of Mechanics*, Benjamin Cummings, Reading, Mass., 1978.

[3] Ceasareo, G. , and Nicolo, F., "DYMIR: A code for generating dynamic models of robots," preprint, 1984.

[4] Cox, B. J., *Object Oriented Programming – An Evolutionary Approach*, Addison-Wesley Publishing Co., Reading, Mass., 1986.

[5] Dillon, S. R., "Computer assisted equation generation in linkage dynamics," Thesis, Ohio State University, 1973.

[6] Fletcher, H. J., Rongved, L., and Yu, E. Y., "Dynamics analysis of a two-body gravitationally oriented satellite," *Bell Systems Tech. J.*, Vol. 42, pp. 2239-2266, 1963.

[7] Frisch, H. P., "A vector-dyadic development of the equations of motion for N-coupled rigid bodies and point masses," *NASA TND-7767*, Oct., 1977.

[8] Frisch, H. P. "The NBOD2 User's and Programmer's Manual," in *NASA TP 1145*, Feb., 1978.

[9] Grossman, R., Krishnaprasad, P. S., and Marsden, J. E., "Dynamics of two coupled three dimensional rigid bodies", *Dynamical Systems approaches to Nonlinear Problems in Systems and Circuits*, F.M.A. Salam and M. Levi, eds., SIAM, Philadelphia, pp. 373–378, 1987.

[10] Hooker, W. W., and Margulies, G., "The dynamical attitude equations for a n-Body Satellite," *J. Astronaut. Sci.*, Vol. 12, pp. 123–128, 1965.

[11] Hunt L. R., Su, R., and Meyer, G., "Global transformations of nonlinear systems," *IEEE Transactions on Automatic Control*, Vol. AC-29, pp. 24–31, Jan. 1983.

[12] Isidori, A., *Non-Linear Control Systems: An Introduction*, Lecture Notes in Control and Information Science, Vol. 72, Springer-Verlag, New York, 1985.

[13] Khalil, W., *Modelization et commande par calculeteurs du manipulateurs*, Thesis, Universitedes Sciences et Technique du Languedoc, 1976.

[14] Krishnaprasad, P. S., "Lie-Poisson structures, dual-spin spacecraft and asymptotic stability," *Nonlinear Analysis, Theory, Methods and Applications*, Vol. 9, No. 10, pp. 1011–1035, 1985.

[15] Liegeois, A., Khalil, W., Dumas, J. M., and Renauld, M., "Mathematical and computer models of interconnected mechan-

ical systems," *2nd International Symposium on Theory and Practice of Robots and Manipulators*, pp. 5–17, Sept. 1976.

[16] Likins, P. W., *Analytical Dynamics and Nonrigid Spacecraft Simulation*, NASA Tech. Rept. 32-1593, July 15, 1974.

[17] Marsden, J. E., and Weinstein, A., "Reduction of symplectic manifolds with symmetry," *Rep. Mathematical Physics*, Vol. 5, pp. 121-130, 1974.

[18] Murray, J. J., and Neuman, C. P., " ARM: An algebraic robot dynamic modeling program," IEEE, 1984.

[19] Quartararo, R., "Two-body control for rapid attitude maneuvers," *Advances in Astronautical Sciences*, L.A. Morine, ed., Vol 42, pp. 397-422, American Astronautical Society, San Diego (AAS 80-203), 1980.

[20] Roberson, R. E., and Wittenburg, J., "A dynamical formalism for an arbitrary number of interconnected rigid bodies with reference to the problem of satellite attitude control," *Proc. of the 3rd Intl. Congress of Automatic Control, London, 1966*, pp. 46D.1-46D.8, Butterworth and Co., Ltd, London, England, 1967.

[21] Rosenthal, R. E., and Sherman, M. A., "High performance multibody simulations via symbolic equation manipulation and Kane's method," *J. of Astronautical Sciences*, Vol. 34, No. 3, pp. 223-239, July-Sept. 1986.

[22] Schwertassek, R., and Roberson, R. E., "A state-space dynamical representation for multibody mechanical systems, Part I: Systems with tree configuration," *Acta Mechanica*, Vol. 50, pp. 141-161, 1984.

[23] Sreenath, N., *Modeling and Control of Multibody Systems*, Ph.D. Thesis, Univ. of Maryland, 1987.

[24] Sreenath, N., and Krishnaprasad, P. S. "DYNAMAN: A tool for manipulator design and analysis," *IEEE Intl. Conf. on Robotics and Automation*, San Francisco, Calif., Vol. 2 , pp. 836–842, April 7-11, 1986.

[25] Sreenath, N., Oh, Y. G., Krishnaprasad, P.S., and Marsden, J. E., "Dynamics of coupled planar rigid bodies Part I : reduction, equilibria and stability," *J. of Dynamics and Control of Systems*, to appear.

[26] Stefik, M., and Bobrow, D. G., "Object-oriented programming: themes and variations," *The AI Magazine*, pp. 40-62, 1986.

[27] Sturges, R., "Teleoperators arm design program," *Report 3.2746*, MIT Draper Lab, Cambridge, Mass., 1973.

[28] Wertz, J., *Spacecraft Attitude Determination and Control*, D. Reidel, Dordrecht, 1978.

[29] Wittenburg, J. M., *Dynamics of Systems of Rigid Bodies*, B.G. Tuebner and Co., Stuttgart, West Germany, 1977.

[30] Wittenburg, J. M., "Dynamics of Multibody Systems," *Theoretical and Applied Mechanics*, Proc. XV IUTAM Conf. Toronto, pp. 197-207, 1980.

[31] Wittenburg, J. M., "MESA-VERDE: A symbolic program for nonlinear articulated-rigid-body dynamics," *ASME Design Engg. Division Conf.*, Cincinnati, Ohio, Sept., pp. 10-13, 1985.

[32] Velman, J. R., "Simulation results for a dual-spin spacecraft," *Proc. of the symposium on Attitude Stabilization and Control of Dual-Spin Spacecraft, El Segundo, Calif., 1967*, Rept. SAMSO-TR-68-191, 1967.